D1410212

Digital Image Processing of Remotely Sensed Data

Notes and Reports in Computer Science and Applied Mathematics

Editor
Werner Rheinboldt
University of Pittsburgh

1. John R. Bourne. Laboratory Minicomputing

2. Carl Tropper. Local Computer Network Technologies

3. Kendall Preston, Jr. and Leonard Uhr (Editors). Multicomputers and Image Processing: Algorithms and Programs

5. R. Michael Hord. Digital Image Processing of Remotely Sensed Data

In preparation

4. Stephen S. Lavenberg (Editor). Computer Performance Modeling Handbook

Digital Image Processing of Remotely Sensed Data

R. Michael Hord
General Research Corporation
Westgate Research Park
McLean, Virginia

1982
ACADEMIC PRESS
A Subsidiary of Harcourt Brace Jovanovich, Publishers
New York London
Paris San Diego San Francisco São Paulo Sydney Tokyo Toronto

ACADEMIC PRESS, INC.
111 Fifth Avenue, New York, New York 10003

United Kingdom Edition published by
ACADEMIC PRESS, INC. (LONDON) LTD.
24/28 Oval Road, London NW1 7DX

Library of Congress Cataloging in Publication Data

Hord, R. Michael, Date
Digital image processing of remotely sensed data.

(Notes and reports in computer science and applied mathematics)
Includes index.
1. Image processing--Digital techniques. 2. Remote sensing. I. Title. II. Series.
TA1632.H67 1982 621.36'78 82-16267
ISBN 0-12-355620-1

PRINTED IN THE UNITED STATES OF AMERICA

82 83 84 85 9 8 7 6 5 4 3 2 1

For Susan

Contents

4
Algorithms

5
Application Examples

6
Research Topics

7
Practical Issues

Preface

Today, hundreds of pictures of the earth were taken. There were hundreds taken yesterday and hundreds will be taken tomorrow. Most of these are converted into arrays of numbers for transmission and processing. Over the past decade, particularly the last five years, a rich technology has emerged to use these arrays of numbers to monitor crops and natural resources, prospect for minerals, forecast weather, assess pollution, help planning of regions and urban areas, route shipping, and, in general, manage the environment. The effective exploitation of these pictures for these and other applications requires an understanding of the remote sensing and processing of the data that carry the information contained in these pictures. A large number of government organizations, private companies, and universities are actively engaged in the digital image processing field and continue to develop these digital exploitation techniques.

In this volume, a practical approach to the subject is presented with emphasis on application examples and algorithms. In this way, the reader can understand the effect of each processing step and, hence, become an informed user of this technology. For those readers with access to the appropriate resources, enough detail is provided to encourage them to try their hand at it: where to get the data and what is available, what preprocessing is needed to prepare the imagery for processing, before and after illustrations for algorithms, and vendors of special-purpose hardware. Research topics are described to indicate the current limits of these computer methods.

The reader who is a professional regional planner or natural resource manager would be expected to benefit from this book, as would hydrologists, foresters, agronomists, geologists, and the like. In general, this book is directed toward those who would like to get started in the doing or using of digital methods to extract information from terrain photos.

The emphasis is on satellite imagery, particularly the imagery produced by the Multispectral Scanner of the Landsat program. Where possible, the use of mathematics has been kept to a minimum, except in Chapter 6, Research Topics.

In most cases, the coverage of the material is characterized by breadth. Of course much has, of necessity, been omitted, but in the treatment of each topic the rule has been to present the overview and provide the reader with references to sources of more detail. In some areas, however, the full depth of the topic is explored. In these cases the reader is led from fundamental concepts through intermediate stages to the state of the art.

Acknowledgments

In largest measure I am indebted for assistance in the preparation of this volume to two individuals, Charles Sheffield of Earth Satellite Corporation and Azriel Rosenfeld of the University of Maryland. They have over several years provided insight and encouragement, and their influence pervades this work.

More generally, Earth Satellite Corporation, through J. Robert Porter, Jr., Richard Anderson, John Everett, Robert Macomber, Mel Erskine, Roger Nagel, and others, provided a majority of the images and material included, and is gratefully acknowledged for substantial cooperation in the preparation of this volume.

Gratitude is also expressed to David Simonett, Robert Haralick, Nicholas Gramenopoulos, Robert Shotwell, James Pottmyer, Archibald Park, William Smith, Kent Bowker, John Gerdes, Lawrence Gambino, Earl Merritt, James Wray, Chuck Poulton, and Wayne Rohde.

A good deal of the material for this book was adapted from government documents too diverse to list here. References to the original public domain publications are generally provided in the text.

Figures 5.3, 5.6 through 5.8, 5.12, 6.15, and 6.16 originally appeared in "Digital Enhancement of Landsat MSS Data for Mineral Exploration," chapter 9 of *Remote-Sensing Applications for Mineral Exploration,* edited by William Smith, Dowden, Hutchinson & Ross publisher, 1977.

1 Summary of Basic Concepts

REMOTE SENSING[1]

Since about 1945, a variety of nonphotographic image-forming systems has been developed to detect radiation reflected or emitted (or both) from a remote scene. These responded to the problems inherent in the use of that most common of optical imaging devices, the camera.

Cameras are small, lightweight, and relatively simple to use. But they have two serious limitations in remote-sensing applications. First, the output is a photograph, which is awkward to telemeter and to process for analysis. Second, photographic film is limited in spectral response to the region from the near ultraviolet to the near infrared. Therefore, night operation is almost impossible unless artificial light sources are used for illumination. Moreover, clouds, fog, and smoke are opaque in this spectral region so that "seeing" the ground from very high altitudes is often difficult and sometimes impossible.

Nonphotographic sensors operate in portions of the electromagnetic spectrum from the microwave to the ultraviolet region. Infrared, passive-microwave, and radar sensors operate under both day and night conditions, and

[1]This section was adapted from Holter, M. R., "Imageing with nonphotographic sensors," *Remote Sensing, with Special Reference to Agriculture and Forestry*. Washington, D.C.: National Academy of Sciences, 1970.

radar sensors are not seriously hindered by clouds and bad weather. Since the sensor data are collected in electrical form, they are easily transmitted to a remote location. In addition, this imagery can be processed with electronic circuits. These nonphotographic sensors, however, are generally more complicated, larger, and have lower spatial resolution than comparable photographic equipment.

The electomagnetic spectrum extends from radio frequencies (fractions of a hertz) to cosmic rays (frequencies exceeding 10^{26} Hz [hertz]). The spectrum is commonly divided approximately as shown in Table 1.1.

Microwave

Passive-microwave and radar imagery, because of their longer wavelengths, cannot have as high a resolution as infrared or photographic images. The theoretical limit in resolving power of an optical system or radar antenna with a circular aperture is $1.22\ \lambda/D$, where λ is the wavelength and D is the diameter of the aperture. To avoid diffraction or sidelobe effects, D should be many times larger than λ. At optical frequencies, this is achieved with apertures of reasonable physical size. Unfortunately, to obtain equivalent ratios at microwave frequencies, the antenna must be extremely large. The exception is the special technique of coherent synthetic-aperture radars.

Until discovery of the synthetic-aperture techniques, obtaining high-resolution imagery at microwave frequencies did not appear promising. The synthetic-aperture radar, in effect, allows effective λ/D ratios to approach those of optical systems while physical antennas are kept down to practical size. As a result, active-microwave imagery resolution approaches that obtained by optical systems. In addition, radar systems operating at wavelengths longer than 3 cm have the great advantage of all-weather, day or night capability and are not limited by clouds. Perhaps most important of all, their resolution is not limited by range, an extremely important factor when the distance between the imaging sensor and the scene is great. In general, however, the sizes, power requirements, and complexity of synthetic-aperture radar systems are far greater than those required for optical-frequency equipment.

In the millimeter wavelength region (.1–1 cm) and in the centimeter wavelength region (1–10 cm), there is sufficient scene-emitted thermal radiation that passive sensing is possible, although not with the same temperature sensitivities as in the infrared region.

Intensive development of radar systems has been underway for four decades, and a wide variety of types exists. One of these types, side-looking

TABLE 1.1
Wavelength and Frequency Ranges of Operation for Remote Sensors

Spectral region	Wavelength	Frequency (Hz)	Common applicable imaging sensors
Microwave (Active [radar] passive)			
Decimeter (UIIF)	10–100 cm	3×10^9 to 3×10^8	
Centimeter	1–10 cm	3×10^{10} to 3×10^9	Scanning antennas with radio-frequency receivers
Millimeter	.1–1 cm	3×10^{11} to 3×10^{10}	
Optical (Infrared)			
Far infrared	8–1000 μm	3.75×10^{13} to 3×10^{11}	Scanners with infrared detectors
Intermediate infrared	3–8 μm	1×10^{14} to 3.75×10^{11}	Various image tubes (not very satisfactory)
Near infrared	.780–3 μm	3.85×10^{14} to 1×10^{14}	Photographic film to approximately 1 μm
			Scanners with infrared detectors
			Various image tubes
Visible	.380–.780 μm	7.89×10^{14} to 3.85×10^{14}	Photographic film
			Scanners with photomultiplier detectors
			Television
Usable			
Near ultraviolet	.315–.380 μm	$.952 \times 10^{15}$ to 7.89×10^{14}	Photographic film (quartz lenses)
			Scanner with photomultiplier detectors
Middle ultraviolet	.280–.315 μm	1.08×10^{15} to $.952 \times 10^{15}$	Image-converter tubes
Far ultraviolet	.010–.280 μm	3.0×10^{15} to 1.08×10^{15}	These wavelengths do not penetrate the earth's atmosphere significantly so are not useful for remote sensing
Vacuum ultraviolet	.004–.010 μm	7.5×10^{14} to 3.0×10^{15}	

airborne radar (SLAR) systems, views at right angles to the aircraft or spacecraft flight line. In that direction, that is, the range direction, scanning is accomplished by measuring the time the emitted pulse takes to travel from the antenna to the ground and back to the antenna. Thus, travel time will differ for objects at different distances. Scanning along the direction of the vehicle flight line is accomplished by the vehicle motion, which causes the position of each transmitted and received pulse to be displaced slightly from the previous one. The usual system has a transmitter, an antenna, a detecting element, and a display. When the synthetic-aperture, resolution-enhancing technique is employed, the display is replaced by a recording mechanism, usually photographic film or magnetic tape. The recorded signals are subjected to a sophisticated processing, and the resulting enhanced image is displayed on a cathode-ray tube (CRT), which is photographed. Present-day transmitter power and receiver sensitivities that can be achieved from aircraft or spacecraft are very high and quite capable of producing imagery of photographic quality.

State of the art passive-microwave sensors are not as well developed as sensors for the other spectral ranges. At passive micro-wavelengths, scene-emitted radiation is much less than in the infrared and, as a consequence, the highest sensitivities and lowest internal noise powers are required of the receivers.

Very often the detector is caused to view alternately an internal calibrated source and the scene to be examined; as a means of overcoming instabilities and providing higher performance, only the difference signal is recorded. This is called the *Dicke radiometer technique*. Detector sensitivities and internal noise are the present limiting factors, and improvements may be expected. Because of the sensitivity limitations, the passive-microwave sensors have the coarsest resolution limitation of any of the sensors discussed.

Infrared

The infrared portion of the electromagnetic spectrum lies between the visible and the microwave regions (.78 μm to 1000 μm). The existence of this "invisible" radiation was discovered in 1800 by Sir Frederich William Herschel, who used a thermometer to detect the energy beyond the red portion of a visible "rainbow" produced by a prism. In 1861, Richard Bunsen and Gustav Kirchhoff established the principles underlying infrared spectroscopy, a powerful tool in the basic studies of molecular structure. After a century of further progress, the uses of infrared technology have been extended to a host of other applications. Among them is the use of infrared sensors in producing images of remote scenes. These remote-sensing systems have been used on the ground, in airplanes, and in space vehicles. Compared with their radar counterparts, these devices are small, lightweight, and require little electric power. Like radar, infrared systems operate under both daytime and nighttime conditions. However, infrared systems are hampered by clouds and rain—a problem radar systems do not have when sufficiently low frequencies are used. The two important features of infrared surveillance systems that give them an advantage over radar are the high spatial resolution achievable with relatively simple designs and the fact that they operate passively; that is, they do not need to illuminate the ground scene artifically in order to record an image of it.

The infrared radiation emanating from a scene is due to both self-emission from each object within the scene and to reflected radiation from the object as a result of illumination from natural sources such as the sun, clouds, moon, stars, and aurorae. Generally speaking, in the daytime the radiation is predominantly reflected in the spectral region from .78 μm to 3.0 μm and is mostly self-emitted at wavelengths longer than 4.5 μm. During daylight hours, roughly comparable amounts of reflected sunlight and emitted radiation will be present in the wavelength region 3 μm–4.5 μm.

A factor that must be considered in the design of infrared systems is the transmission of the atmosphere. At infrared wavelengths, molecular absorption has its own characteristic infrared-absorption spectrum. The gases that have the most significant effect include CO_2, N_2O, H_2O, and O_3. For those portions of the infrared spectrum where one or more of these gases is strongly absorbing, the atmosphere is essentially opaque. The spectral intervals at which the atmosphere is reasonably transparent are (approximately, in microns) 7–1.35, 1.35–1.8, 2.0–2.4, 3.5–4.1, 4.5–5.5, 8–14, 16–21, and 750–1000.

Infrared images can be obtained with a number of devices: infrared photographic film, image tubes, and optical scanners. Infrared photographic film is used in the same way as film in conventional photography. It is made by adding sensitizing dyes during the manufacturing process to extend the spectral response to about 1.2 μm. This range is adequate for sensing vegetation condition. The reflectance spectrum of green plants ranges in value from about 5% at .6 μm to about 45% at .75 μm; this region, which contains a pronounced chlorophyll absorption band, is very sensitive to changes in the health of vegetation.

Image tubes are electronically scanned image-forming devices such as those used in television systems. No mechanical moving parts are required, although some of them use an electron beam to convert the invisible infrared image on the sensitive surface to a visible image on a phosphor screen. Infrared image tubes that are sensitive to wavelengths longer than 1.2 μm tend to have long time constants; that is, the speed at which an image is formed is quite slow. Although this sluggish response is not a severe problem for some groundbased systems, it is a problem for airborne or space applications and therefore will not be discussed further here.

The most widely used airborne infrared surveillance device is the scanner. In systems of this kind, a single-element detector, or a one- or two-dimensional array (*mosaic*) is used to view small parts of the total field of view. This type of sampling is generally called *optical-mechanical scanning* or simply *scanning,* as opposed to the electron-beam scanning associated with image tubes. There are, in principle, two types of scanning—image plane and object plane. The first type is so named because an image of the total object field is formed by a collecting telescope. This image is then sampled by a second optical system containing the infrared detector. In this case, the scanning device can be small and the optics for it easily designed. However, a high-quality image of the object field must be formed by the telescope. Designing and constructing the optics for such a telescope when the field of view is larger than 10° or 20° is a difficult problem.

The object-plane scanner generally consists of a rotating mirror and a telescope that focuses the radiation from a small portion of the object plane. The rotating mirror causes the field of view to move smoothly across the object plane. When such a system is placed in an airplane, a strip recording of the image of the ground scene is produced. Such devices have a very fine angular resolution (~1 mrad^2 [milliradian]). This means an instantaneous ground resolution of 1 ft^2 (.09 m^2) for an aircraft altitude of 1000 ft (300 m). For satellites at altitudes of 100 miles, a scanner with 1 mrad^2 angular resolution would view an instantaneous ground patch of 600 × 600 ft (184 m × 184 m).

Consider a surveillance aircraft that carries a scanner having an instantaneous field of view (resolution element) of angular size B in both dimensions. To provide a good-quality image, B is typically quite small, on the order of milliradians. This instantaneous field of view is caused to scan through an angle at right angles to the aircraft path by a rotating element in the scanner. The motion of the aircraft carries the scanner forward so that successive scans cover different strips of the ground. The portion of ground swept during a single scan is called a *line.* If successive lines are not contiguous (if parts of the ground between the lines are not scanned), the condition is termed *underlap.* This occurs if the aircraft speed is too high or if the rotating elements revolve too slowly. If successive lines scan partly over the same terrain, the condition is called *overlap*. Underlap is undesirable, since information is not obtained between successive lines. Contiguous scanning is desirable because no ground remains unscanned and there is a certain economy in not scanning more than once over any part of the ground. Under certain circumstances, overlap is useful since the redundancies can be used to increase the signal:noise ratio.

Visible

Photographic and television devices are the most common types of image sensors used in the visible spectrum. However, where multispectral information both within and outside the visible region is required and extensive ground areas are to be studied from unmanned spacecraft, neither photographic nor television devices are optimum. To use several visible-region sensors in conjunction with infrared sensors in a multispectral system, all must view an identical scene at the same instant and have the same dwell time and observation geometry. This can best be done if all the sensors have the same system configurations. Since photographic films are not sensitive to wavelengths longer than about 1 μm, the infrared devices are for the most part electro-optical line scanners. Thus, the visible sensors are electro-optical line scanners as well.

In the visible region of the electromagnetic spectrum, the irradiance at a remote sensor is determined primarily by the characteristics of the illuminating source, the reflectance of objects, the spatial distribution of the objects, and the atmospheric transmission. The total natural illumination on an object at the earth's surface is the sum of the direct illumination, aerosol scattering, and the reflections from other objects. It varies greatly with type and amount of cloud cover, haze, and other environmental factors.

Another important factor that affects the visible radiation received at the remote sensor is the object's reflectance. The incident natural illumination has a spectral distribution and since the reflectance also varies with wavelength, the wavelength variation of the reflected energy is described by the product of these spectra. The spectral distribution of the light arriving at the sensor is further changed by the transmission of the atmosphere.

The reflectance function varies considerably from one object to another. This variation is caused by a number of factors, including, in the case of vegetation, level of maturation, soil condition, and thermal history of the environment; for other objects, the variation is caused in part by differences in the surface topography and the moisture content.

The attenuation of visible-region electromagnetic radiation through the atmosphere depends on absorption due to gasses and water vapor and on scattering by atmospheric aerosols. Attenuation by absorption plays a minor role in this spectral region. An absorption band for molecular oxygen and water vapor exists at about .68 μm. This band can attenuate 10% or more for a vertical path through the atmosphere to satellite-borne sensors.

Attenuation due to scattering of radiation is the most significant factor in the visible spectrum. It is caused by an interaction between the radiation and small particles that range in size from less than 1 μm to as large as 10 μm (water droplets). The amount of scattering will vary with the size of the scattering particle and the wavelength of the radiation being scattered. These parameters have three commonly recognized ranges. When the diameter of the scattering particles is significantly less than the wavelength of the radiation, the scattering is proportional to $1/\lambda^4$, where λ is the wavelength. This is called *Rayleigh scattering*. When the diameter of the scattering particle is of the order of the wavelength being scattered, a theory formulated by Mie must be considered. In this region the amount of scattering will be a varying function of λ.

The third commonly recognized scattering region, that is, when the scattering particles are significantly greater in diameter than λ, there is no wavelength dependence of the scattering. An example of this is the white appearance of scattering from fogs of large-diameter water particles. Water droplets in the form of haze, fog, clouds, and dust particles constitute an aerosol that scatters radiation significantly in the visible and ultraviolet portions of the spectrum. A substantial amount of data on aerosol scattering exists. However, the problem is complex, and reasonable, predictive, analytical transmission models for other than clear days have not been developed. In general, one can say that the earth as viewed from a satellite has 50% of its surface area obscured by clouds at any specific time.

Except for the detector element, line scanners for use in the visible region are identical in design to those used in the infrared region. The detectors used in scanners for the visible spectrum differ considerably from those used in the infrared. In the visible region they are often photoemissive types. Here, incident photons cause electrons to be emitted by the photosensitive surfaces. These electrons are then collected by an anode, and the electric current through the circuit is proportional to the incident radiation intensity.

In addition to the photoemissive detectors, the infrared photoconductive detectors are also sensitive to visible electromagnetic radiation and are often used in line scanners with spectral filters that limit the incident radiation wavelengths to those desired. Their sensitivity in the visible region, however, is not as great as that of photomultipliers.

DIGITAL IMAGERY

The image or picture that is produced by a camera or nonphotographic imaging sensor is a flat scene representation that varies in visual properties from point to point. The varying visual properties include brightness, color, or reflectance. This variation can be described mathematically by a function of two variables. When color is involved, the function should be regarded as vector valued or else several functions should be used. Each of these several functions or the single function for a monochromatic (black and white) picture is real valued, nonnegative, bounded, and analytically well behaved. The value at a point of this function is termed the *gray level.*

When a picture is digitized it is represented by a regularly spaced array of samples of this picture function. These samples are quantized so the numbers take on a discrete set of possible values, generally integers. The elements of a digital picture array are called *pixels.* If just two gray levels, "black" and "white", are used, they are represented by 0 and 1, and the picture is called a *binary image.*

The picture arrays that are processed in practice can be quite large. To adequately represent an ordinary television image digitally, an array of about 500 lines with about 500 samples per line is required. Each pixel can take on about 40 gray levels. For computer processing, these gray levels are coded as binary numbers; to allow a 40 gray-level dynamic range, a 6-bit number is required for each pixel. Hence a single still television picture needs about 1.5 million bits. A Landsat multispectral image (see p. 22) uses about 250 million bits.

Digital image processing encompasses a wide variety of techniques and mathematical tools. They have all been developed for use in one or the other of the two basic activities that constitute digital image processing, image enhancement, and image analysis. *Image enhancement* accepts a digital image as input and produces an enhanced image as output; in this context *enhanced* means better in some respect. This includes increasing or decreasing the contrast, removing or introducing geometric distortion, edge sharpening or smoothing, or altering the image in some respect that facilitates the interpretation of its information content. *Image analysis,* on the other hand, accepts digital imagery as input but as output produces a report of some type. It may be a simple count of the number of triangles in the format or a complex report that assigns each pixel to a land cover category and a tabular list of the aggregate acreage of each land cover category in each census tract in the image.

Because digital image processing does encompass such a wide variety of techniques and mathematical tools, each application involves a sequence of processing steps. These steps are modules that can be arranged in an enormous number of different sequences, and each module generally admits the specification of parameter values that further expand the range of possible manipulation schemes to achieve the desired results. The major advantages of digital images as compared with analog (e.g. film) images are the quality of the data and the accessibility of that data to computer manipulation. The quality advantage is really an aggregate of two advantages:

1. *Dynamic range:* Up to 128 or more gray levels can be present on digital images whereas only 15–30 are discernible on film products.
2. *Repeatability:* A copy of a copy of a copy of a digital image contains exactly the same data as the original, whereas a fourth-generation photographic product is substantially degraded in comparison with the original.

Although these are strong advantages, the compelling advantage for the applications community derives from the accessibility of the data to computer manipulation. To generate a general purpose film product is essentially to choose a compromise among competing requirements from agronomists, hydrologists, foresters, geologists, urban planners, and other user groups; it is not optimized for any one. Digital images lend themselves to custom tailoring by the user for his or her application in a way that film products cannot.

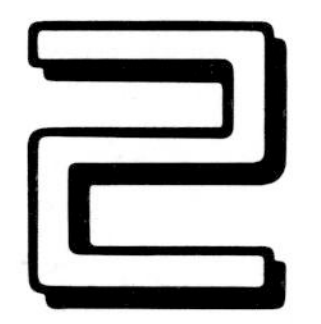

Data Sources

IMAGERY AVAILABLE IN DIGITAL FORMAT

EROS Data Center

LANDSAT DATA

The first Earth Resources Technology Satellite, ERTS–1 (later renamed Landsat–1), was launched on 23 July 1972. Landsat–2 was launched on 22 January 1975. When Landsat–1 expired, Landsat–3 was orbited (1978). Each Landsat flies in a circular orbit 920 km (570 miles) above the Earth's surface and circles the Earth every 103 min, or roughly 14 times per day. Each daytime orbital pass is from north to south. From such a vantage point, each Landsat can cover the entire globe, except for the poles, with repetitive coverage every 18 days. Each of the satellites views the Earth at the same local time, roughly 9:30 a.m. at the equator, on each pass. The sensors on board the spacecraft transmit data to NASA ground receiving stations in Alaska, California, and Maryland either directly or from tape recorders. The data are converted from electronic signals to photographic images and computer compatible tapes at NASA's Goddard Space Flight Center in Greenbelt,

Maryland. Master reproducible copies are flown to the EROS Data Center (EDC) in Sioux Falls, South Dakota, where images are placed on file and where requests for reproductions are filled for the scientific community, industry, and the public at large. Because of the experimental nature of the satellites and the limited capabilities of NASA ground-processing equipment at Greenbelt, approximately 30 days or more are required from the time the signals are first received on the ground to the time that the data are available to the public at the EROS Data Center.

Each Landsat presently carries two data acquisition systems: (*a*) a multispectral scanner and (*b*) a return beam vidicon (RBV) or television system. The multispectral scanner, or MSS, is the primary sensor system and acquires images of 185 km (115 miles) per side in four spectral bands in the visible and near-infrared portions of the electromagnetic spectrum. These four bands are

Band 4, the green band, .5–.6 μm, emphasizes movement of sediment-laden water and delineates areas of shallow water, such as shoals, and reefs.

Band 5, the red band .6–.7 μm, emphasizes cultural features, such as metropolitan areas.

Band 6, the near-infrared band, .7–.8 μm, emphasizes vegetation, the boundary between land and water, and landforms.

Band 7, the second near-infrared band, .8–1.1 μm, provides the best penetration of atmospheric haze and also emphasizes vegetation, the boundary between land and water, and landforms.

An analysis of the four individual black and white images or a false-color composite image often permits users to identify and inventory different environmental phenomena, such as distribution and general type of vegetation, regional geologic structures, and area extent of surface water. The repetitive (9 or 18 days) and seasonal coverage provided by Landsat imagery is an important tool for the interpretation of dynamic phenomena. It should be noted that because of the Earth's rotation and the fact that the image is created by an optical-mechanical scanner, Landsat MSS images are parallelograms, not squares. The sides are parallel to the orbital track of the satellite on the Earth's surface. RBV images have a square format because the image is acquired instantaneously.

The arbitrary forward overlap between consecutive Landsat images is approximately 10%. The sidelap between adjacent orbits ranges from 14% at the equator to 85% at the 80° parallels of latitude.

Latitude and longitude tick marks are depicted on film products at 30′ intervals outside the image edge. These geographic reference marks are annotated in degrees, minutes, and compass direction. A 15-step gray-scale tablet is exposed on every frame of Landsat imagery as it is produced. This scale is used to monitor and control printing and processing functions and to provide a reference for analysis related to a particular image. The annotation block directly over the gray scale contains data that reflect the unique image identification, the geographic location of the scene, and data relative to the existing parameters at the time the data were obtained.

When selecting a single black and white image, it is best to choose band 5. This red band usually gives the best general purpose view of the Earth's surface. A complete set of black and white images from all four bands displays the differences in the appearance of the same area when filtered to green, red, and near-infrared wavelengths. MSS false-color composites are available as standard products. An MSS false-color composite image is generally created by exposing three of the four black and white bands through different color filters onto color film. On these false-color images, healthy vegetation appears bright red rather than green; clear water appears black; sediment-laden water is powder blue; and urban centers often appear blue or blue-gray. MSS false-color composite images that have not already been prepared can be ordered from the Data Center but carry a one-time initial preparation charge, not including the cost of any products ordered from the resulting composite.

A set of Landsat images has been prepared for the coterminous United States. The 470 scenes required to cover the 48 states are available in a single black and white band (band 5), all four bands of black and white, or color composites. The scenes selected were chosen on the basis of quality, time of year (generally spring or summer), and minimum cloud cover.

Landsat data in digital form are available as Computer Compatible Tapes (CCT). The tapes are standard 12.7-mm wide (½ in.) magnetic tapes and may be requested in either 7- or 9-track format at 800 or 1600 bpi (bits per inch). The number of CCTs required (from one to four) for the digital data corresponding to one Landsat scene is dependent on the format requested. In the original MSS tape format, the data for the four MSS bands are interleaved among the tape(s), thereby necessitating all tapes to complete a set. A user profile versus time for Landsat CCTs is shown in Table 2.1. The original Landsat CCT format is described in Figures 2.1, 2.2 and 2.3.

Beginning in 1979, a variety of new CCT formats became available incorporating changes resulting from the new digital image processing systems at NASA's Goddard Center and EDC. Major changes in the format are the addition of ancillary data and that, for the first time, fully corrected scenes (both radiometrically and geometrically corrected) are available. Furthermore, the original band-interleaved-by-pixel-pairs (BIP2) format of the image

TABLE 2.1
User Profile of Landsat MSS CCTs

Category	Scenes (FY 1975)	Scenes (FY 1976)	Scenes (TQ 1976)	Scenes (FY 1977)	Scenes October 1977–August 1978
Federal government (less NASA investigators)	162 (22%)	563 (24%)	214 (21%)	383 (20%)	910 (36%)
NASA investigators	19 (3%)	483 (21%)	128 (13%)	177 (9%)	36 (1%)
State–local governments	48 (7%)	3 (0%)	1 (0%)	29 (2%)	49 (2%)
Academic	182 (25%)	272 (12%)	59 (6%)	182 (10%)	236 (9%)
Industrial	195 (27%)	522 (23%)	235 (23%)	611 (32%)	736 (29%)
Individuals	11 (1%)	38 (2%)	9 (1%)	16 (1%)	13 (1%)
Non United States	108 (15%)	403 (18%)	363 (36%)	489 (26%)	543 (22%)
Other	4 (0%)	5 (0%)	1 (0%)	0 (0%)	0 (0%)
Totals	729 (100%)	2289 (100%)	1010 (100%)	1887 (100%)	2523 (100%)

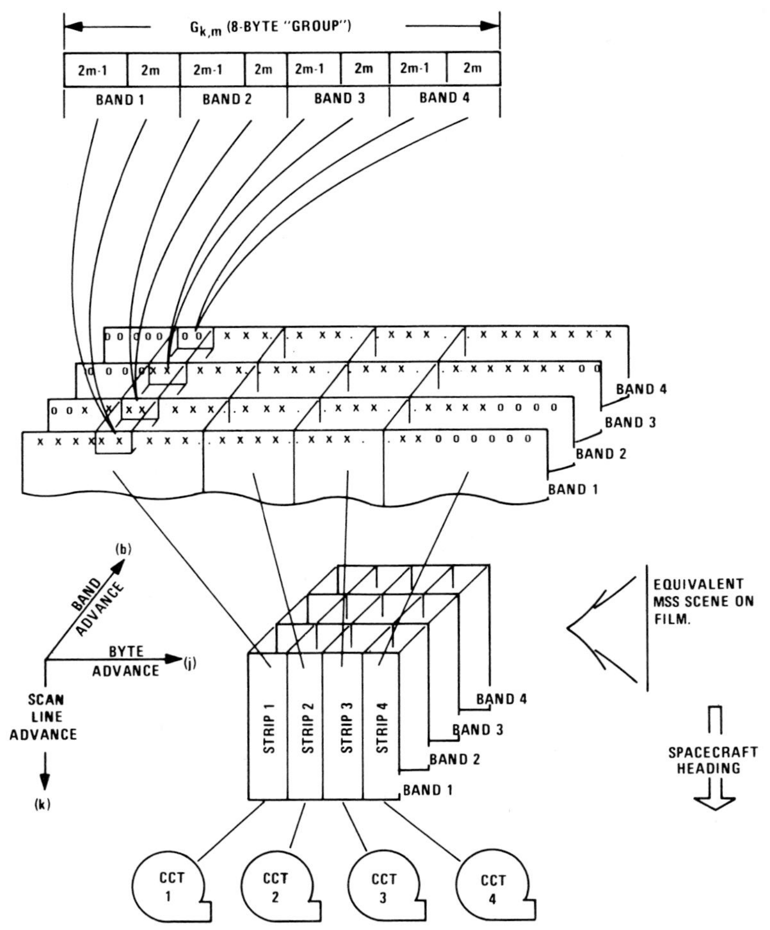

FIGURE 2.1. Bulk MSS 4-band scene to interleaved CCT conversion. O = "dummy" byte; x = "data" byte; k = "scan line" index (there are 4086 scan lines in each strip); m = "group" index (there are 435 groups in each 25 nautical mile interleaved scan line).

data acquired after that date changed to a band-sequential format (BSQ) or band-interleaved-by-line format (BIL).

The set of new formats in which Landsat image data is offered provides the user with a wide range of options. One of the options is geometric correction. Tapes can be ordered without geometric correction or with geometric corrections applied and resampled to a map projection. If geometric correction is

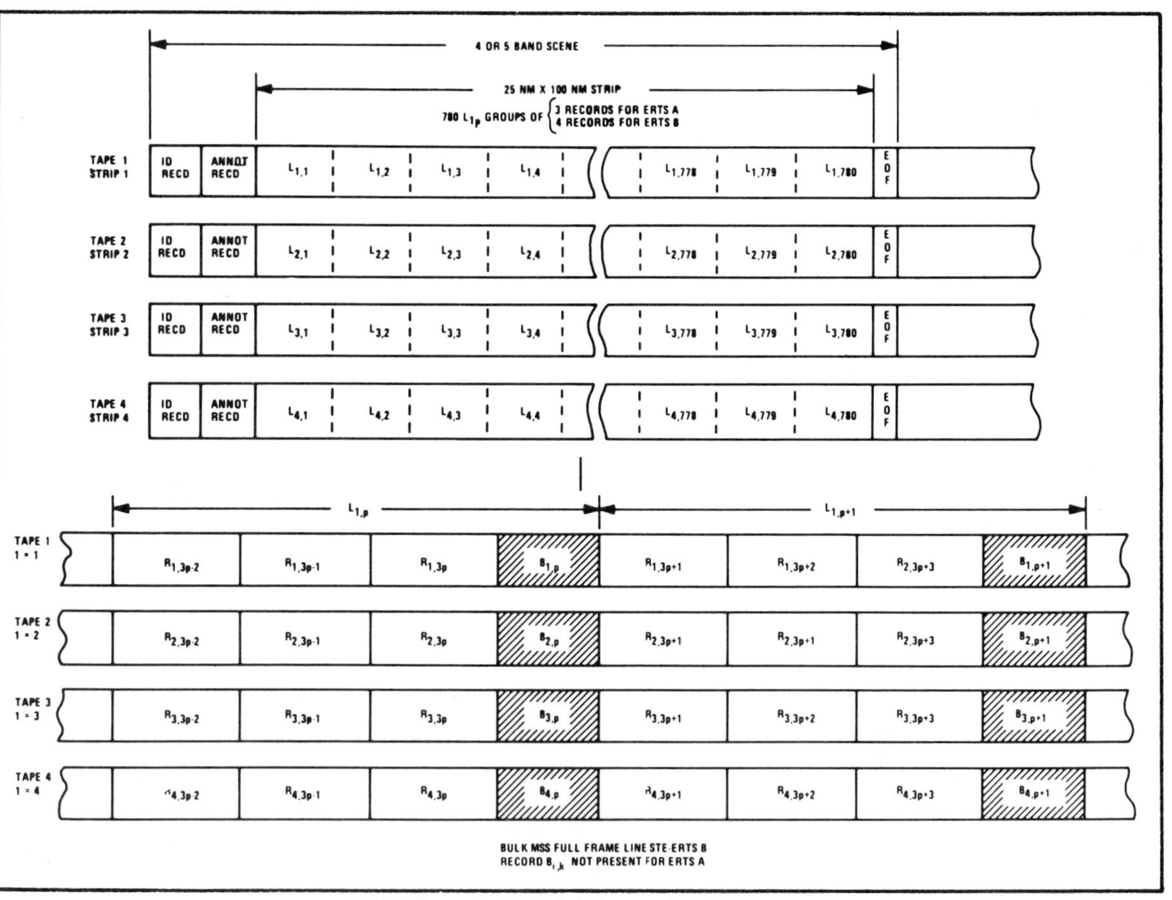

FIGURE 2.2. Bulk MSS full scene, four CTT format.

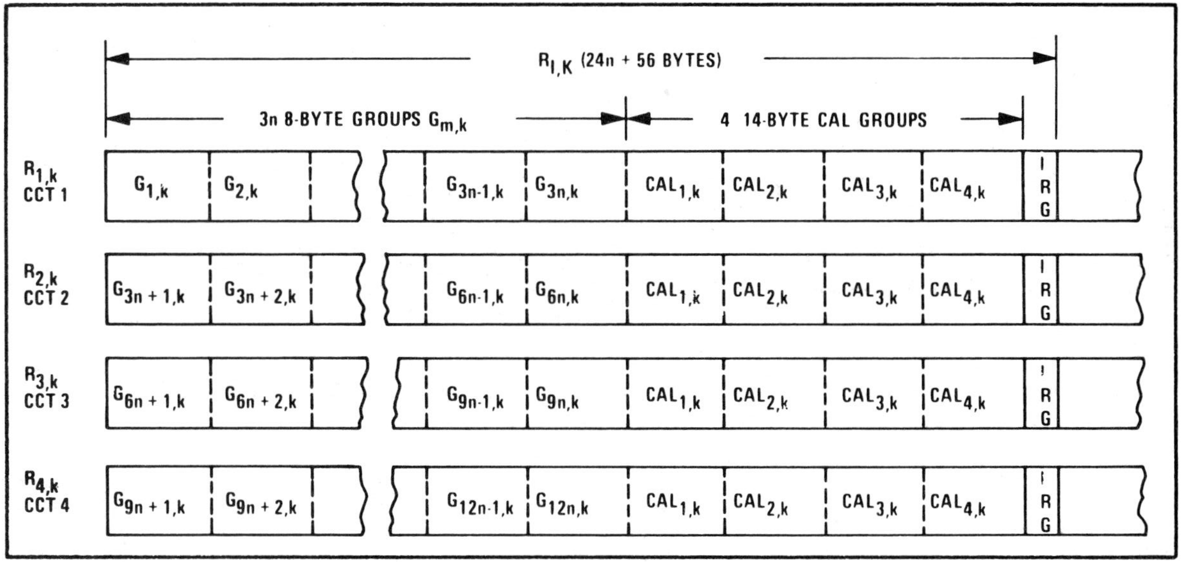

FIGURE 2.3. Bulk MSS full scene interleaved record format.

desired, the user must specify which of three map projections is to be employed: Space Oblique Mercator, Polar Stereographic, or Universal Transverse Mercator. Geometrically corrected tapes have also received various systematic corrections. Included among these systematic corrections are band-to-band offsets, line lengths, earth rotation (skew), and detector-to-detector sampling delay. Some others are pitch, yaw, and altitude. Uncorrected MSS scenes have 3240 pixels per line and 2400 lines, whereas the corrected scenes have 3548 pixels per line and 2983 lines.

Another option available to the user is data enhancement. Contrast enhancement, edge enhancement, and atmospheric scatter compensation are all available and can be applied to geometrically corrected tapes. The contrast enhancement algorithm is either controlled by user-supplied parameters or adapted to the image data. The edge-enhancement algorithm and the atmospheric scatter compensation algorithm employ user-supplied parameter values as well.

With all these and other options available, it is impractical to describe the new formats in detail here. A 75-page document entitled *Manual on Characteristics of Landsat Computer-Compatible Tapes Produced by the EROS Data Center Digital Image Processing System,* for sale by the U.S. Government Printing Office (Stock Number 024-001-03116-7), explains these formats entirely. The original Landsat format described above is for now the format used for the majority of scenes on file, but gradually this will change since all new scenes will be prepared in the new formats.

Figure 2.4 shows an artist's concept of the Landsat vehicle. Figure 2.5 is a photograph of a Landsat satellite at final assembly being prepared in the General Electric Pennsylvania facility for transportation to the launch complex. Figure 2.6 shows a photograph of a typical dish antenna for the ground receiving of satellite telemetered data. Figure 2.7 is a representative Landsat MSS image showing Monterey Bay, California.

After more than 5½ years, during which it orbited the Earth 28,854 times and sent back 271,786 multispectral images, Landsat–1, NASA's first Earth resources monitoring satellite, was turned off 23 March 1978. On 5 March, a more advanced Landsat–3 was launched into a near-polar orbit from the Western Test Range (WTR), Vandenberg Air Force Base, California. In addition to the MSS carried by Landsats–1 and –2, Landsat–3 carried a two-camera panchromatic RBV system, that has transmitted thousands of scenes with a resolution better than 40 m (130 ft).

Also on Landsat–3, a fifth MSS spectral band was provided. The data collected in this thermal-infrared band (10.4–12.6 μm) is characterized by reduced spatial resolution compared with the other four MSS bands. Each

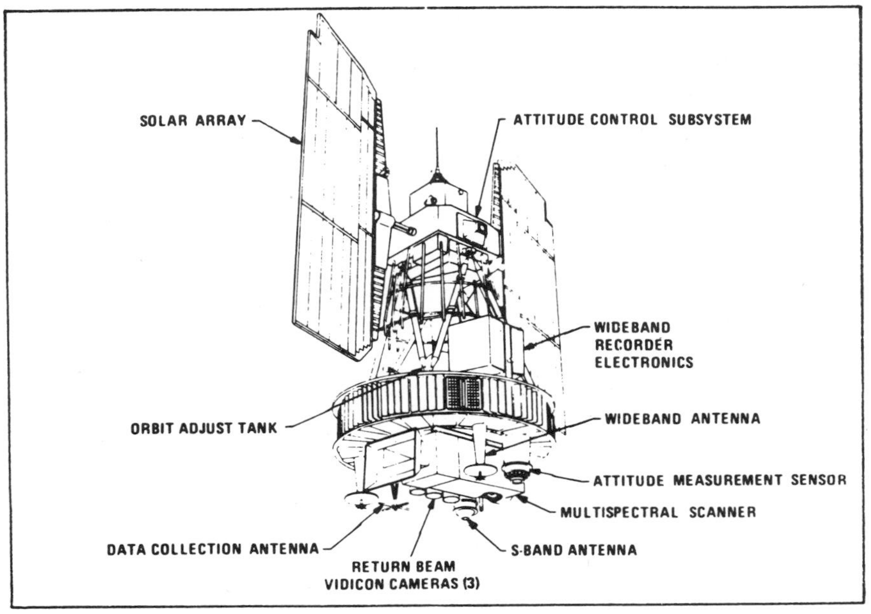

FIGURE 2.4. An artist's concept of the Landsat vehicle.

pixel is replicated by the Goddard Space Flight Center data-processing facility as a three pixel by three pixel block to assure geometric congruence with the other spectral channels. This expanded image array is resampled and corrected to produce the standard thermal image. Thus, the replication pattern is not discernable in a fully processed scene.

Active Landsat receiving stations are located in Brazil, Canada, Italy, Sweden, and the United States. Landsat stations are also operating in Argentina, Australia, Japan, and India. Two active international stations, at Fucino, Italy and Kiruna, Sweden, have been included in the operations of Earthnet, a new program created by the European Space Agency for acquisition, preprocessing, archiving, and distribution of remote sensing data. Its stated purpose is to promote a strong, well-structured user community in Europe. Participating countries include Belgium, France, Germany, Ireland, Italy, Spain, Sweden, and the United Kingdom. Although the central distributing facility is at Frascati, Italy, near Rome, all participating countries have their own National Point of Contact.

Information about foreign Landsat holdings is now available from the EROS Data Center. Nearly 14,000 records from Brazil and 70,000 records

FIGURE 2.5. A photograph of a Landsat satellite at final assembly being prepared in the General Electric Pennsylvania facility for transportation to the launch complex.

FIGURE 2.6. A photograph of a typical dish antenna for the ground receiving of satellite telemetered data.

FIGURE 2.7. A representative Landsat MSS image showing Monterey Bay, California.

from Italy have been added to the EDC data base. The integrated database will provide an indication of what Landsat data are available from the various receiving stations. Orders for data must be directed to the appropriate sites.

The following list of addresses may be used to order data from active international Landsat stations. It should be noted that if you are within a member country of Earthnet, you should order European data from your National Point of Contact; however, if you are anywhere outside the area, you must order from the Earthnet central distributing facility at Frascati, Italy.

Before moving on to other types of satellite imagery, a description of Landsat–D is appropriate. Scheduled for launch in the summer of 1982,

1. User Services Section
   EROS Data Center
   U.S. Geological Survey
   Sioux Falls, South Dakota 57198

2. Instituto de Pesquisas spacias (INPE)
   Departamento de Producao de Imagens
   ATUS-Banco de Imagens Terrestres
   Rodovia Presidente Dutra, Km 210
   Cachoeira Paulista-CEP 12.630
   Sao Paulo, Brazil

3. Canadian Centre for Remote Sensing (CCRS)
   User Assistance and Marketing Unit
   717 Belfast Road
   Ottawa, Ontario K1A 0Y7
   Canada

4. European Space Agency (ESA)
   Earthnet User Services
   Via Galileo Galilei
   000 44 Frascati, Italy

5. Remote Sensing Technology Center (RETEC)
   7-15-17 Roppongi, Minato-Ku
   Tokyo 106, Japan

6. Director, National Remote Sensing Agency
   No. 4 Sardar Patel Road
   Hyderabad-500 003
   Andhra Pradesh, India

7. Australian Landsat Station
   14-16 Oatley Court
   P.O. Box 28
   Belconnen, A.C.T. 2616
   Australia

8. Comision Nacional de Investigaciones Espaciales (CNIE)
   Centro de Procesamiento
   Dorrego 4010
   (1425) Buenos Aires, Argentina

9. Director, National Institute for Telecommunications Research
   ATTN: Stellite Remote Sensing Centre
   P.O. Box 3718
   Johannesburg 2000
   Republic of South Africa

10. Remote Sensing Division
    National Research Council
    Bangkok 9
    Thailand

11. Academia Sinica
    Landsat Ground Station
    Beijing
    People's Republic of China

Landsat–D is the fourth in the Landsat series of experimental satellites designed to explore the Earth's resources from orbit. In addition to the MSS carried by the first three Landsats, Landsat–D will carry a sensor known as the *thematic mapper* (TM), which will provide a spatial resolution approximately three times as detailed as its predecessor MSS.

The new sensor is able to discriminate area features as small as .2 acres (.1 ha) IFOV as compared to the 1.2-acre (.6 ha) IFOV resolution of the earlier systems. (*IFOV* is a technical designation for instantaneous field of view.) Improved spatial resolution, along with the narrower, task-oriented spectral bands, will enable the users to extract much more detailed and timely information.

The thematic mapper spectral bands are as follows:

Band number	Wavelength range (μm)
1	.45–.52
2	.52–.60
3	.63–.69
4	.76–.90
5	1.55–1.75
6	10.40–12.50
7	2.08–2.35

HCMM DATA

The Heat Capacity Mapping Mission (HCMM, April 1979) was the first of a planned series of Applications Explorer Missions. The chief sensor was the Heat Capacity Mapping Radiometer. The objectives included: (*a*) to produce thermal maps at the optimum times for making thermal inertia studies for discrimination of rock types and mineral resources location (Holmes, 1980); (*b*) to measure plant canopy temperatures at frequent intervals to determine the transpiration of water and plant life; (*c*) to measure soil moisture effects by observing the temperature cycle of soils; (*d*) to map thermal effluents, both natural and man made; (*e*) to investigate the feasibility of geothermal source location by remote sensing; and (*f*) to provide frequent coverage of snow fields for water runoff prediction. The high thermal resolution is also used to map thermal gradients in bodies of water.

The radiometer has an instantaneous field of view measuring less than 1 × 1 mrad, high radiometric accuracy, and wide swath coverage on the ground, observed repetitively with a 12-hr period. An infrared band (10.5–12.5 μm) and a Visible band (.8–1.1 μm) are provided. Data tapes are available from the EROS Data Center.

CZCS DATA

The Coastal Zone Color Scanner (CZCS) (NIMBUS–7) Experiment, October 1978, was designed to map chlorophyll concentration in water sediment distribution, gelbstoffe concentrations as a salinity indicator, and temperature of coastal waters and ocean currents (Hovis *et al.*, 1980; Traganza *et al.*, 1980). Reflected solar energy is measured in six channels: .44, .52, .55, .67, .75, and 11.5 μm. Data tapes are available from EDC.

Satellite Data Services

The Satellite Data Services Branch (SDSB) of the National Climatic Center provides environmental and Earth resources satellite data to other users once the priority collection purposes (i.e., weather forecasting) have been satisfied. SDSB is collocated with the operations center of NOAA's National Environment Satellite Service (NESS). NESS manages the national operational environmental satellite program.

DATA HOLDINGS

SDSB files contain data from the early TIROS (Television Infrared Observational Satellite) series of experimental spacecraft in the 1960s; much of the imagery gathered by spacecraft of the NASA experimental NIMBUS series; full-Earth disc photographs from NASA's Applications Technology Satellites I and III, geostationary research spacecraft; tens of thousands of images from the original ESSA and Current NOAA series of Improved TIROS Operational Satellites (ITOS); limited amounts of SEASAT data and both full disc and sectorized images from the Synchonous Meteorological Satellites (SMS) 1 and 2, the operational geostationary spacecraft (GOES). In addition to visible light imagery, infrared data are available from the NIMBUS, NOAA, and SMS satellites. Photographs of the ITOS and SMS satellites are shown in Figures 2.8*a* and 2.8*b*. Each day, SDSB receives about 239 frames from the polar-orbiting NOAA spacecraft, more than 235 SMS–1 and –2 frames, and several special film items. Data from many of the satellites also are available in digital form on magnetic tape.

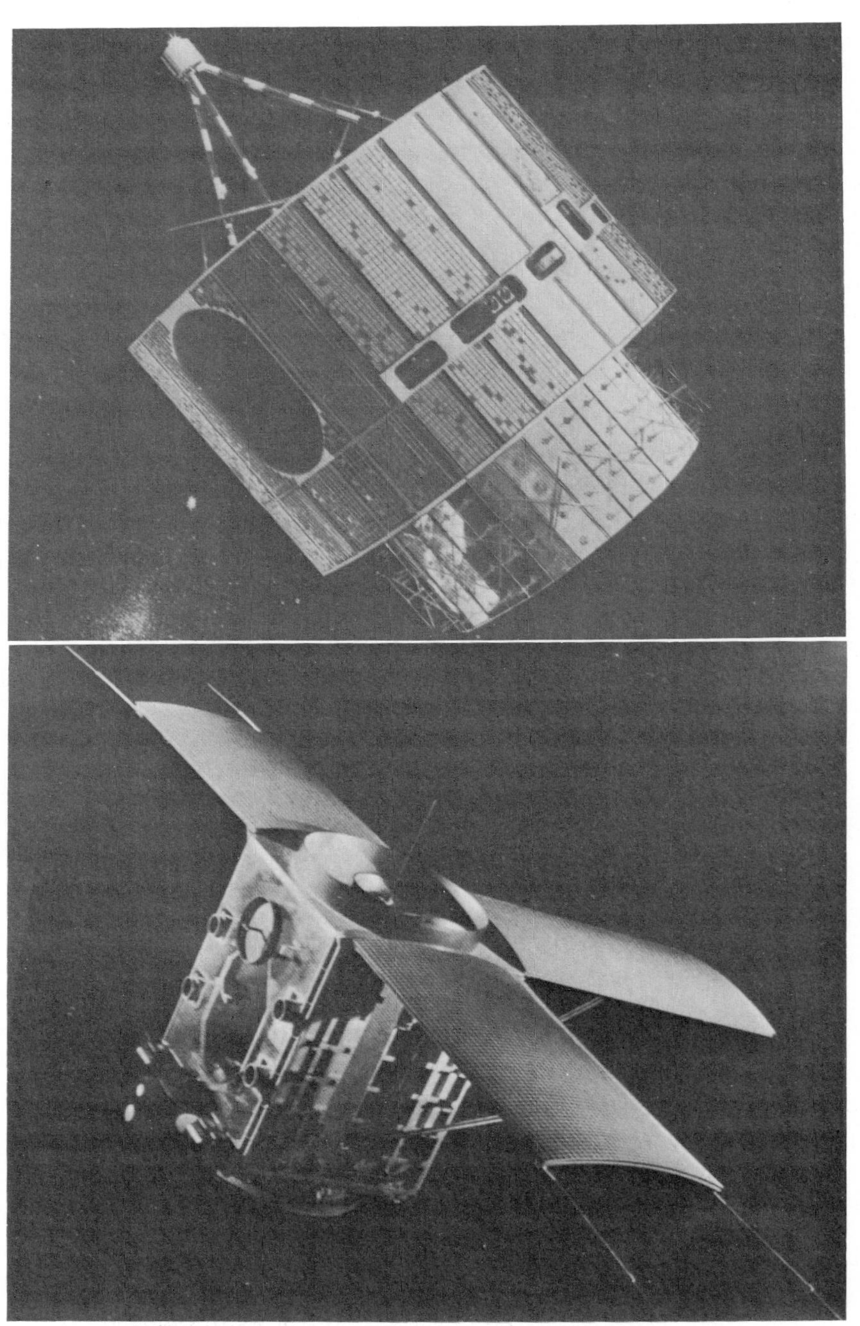

FIGURE 2.8. (*a* and *b*). Photographs of the ITOS and SMS satellites.

USER SERVICES, PRODUCTS, AND PRICES

Potential users can obtain the data they need or more information on the types of data available by calling the Satellite Data Services Branch or by writing to: Satellite Data Services Branch, World Weather Building, Room 606, Washington, D.C. 20233. In addition, several satellite data catalogs are available for reference.

Catalogs. The National Environmental Satellite Service has published a Catalog of Operational Satellite Products that summarizes various photographic and digital data products that can be provided. The catalog is available in limited quantities from SDSB. Also available is a brief explanation of user charges for the products described and information concerning the method of payment.

The Environmental Data Service publishes *Key to Meteorlogical Records Documentation No. 5.4 (Environmental Satellite Imagery),* a catalog issued monthly and describing data available from the NOAA series of operational polar-orbiting satellites. The catalog includes black and white photographs (5 in. (12.6 cm) in diameter) of daily imagery mosaics (visible and infrared) for both the northern and southern hemispheres, and is sold by the National Technical Information Service, U.S. Department of Commerce, Sills Building, 5285 Port Royal Road, Springfield, Virginia 22151.

Additional information may be obtained by writing Support Services, NASA/Goddard Space Flight Center, Code 563, Greenbelt, Maryland 20771; or National Space Science Data Center, Goddard Space Flight Center, Code 601, Greenbelt, Maryland 20771.

Browse Files. To increase public access to Landsat and Skylab imagery, NOAA and the Department of the Interior have established 39 browse files through the United States, located mainly in larger cities. Current addresses and telephone numbers can be obtained from the Satellite Data Services Branch.

Ordering Data. When ordering environmental satellite data, the requester should furnish as much of the following information as possible:

- Satellite from which data are requested
- Date and time of data requested
- Type of data needed (visible or infrared)
- Data format desired (print, transparency, 35 mm film, etc.)
- Use that will be made of the data (to be sure you get what you need)
- Task number to be charged (federal agency) or the name of person or organization to be billed (if nonfederal)

- Address where data are to be sent (plus telephone number)
- Other information that might help SDSB personnel identify and locate the correct data from the archives

Examples of VISSR, VHRR, and HRIR imagery and documentation follow.

EXAMPLE 1. VISSR

The Satellite Data Services Branch (SDSB) of the National Climatic Center has, in conjunction with the National Environmental Satellite Service, established a digital tape archive of SMS–GOES VISSR data. VISSR is an acronym for Visible Infrared Spin Scan Radiometer, an instrument flown aboard the SMS–GOES series of geostationary satellites. For the present, this archive is limited to a set of six 4-mile (6.4 km) resolution sectors per day per satellite, five infrared, and one Visible. Coverage commences with data from August 1976 and includes data from SMS–1, SMS–2, GOES–1, and GOES–2.

The VISSR scanning system consists of a mirror that is stepped to provide north–south viewing while the rotation of the satellite provides west–east scanning. The mirror is stepped following each west–east scan; 1821 scans constitute one full frame, providing full disc Earth coverage. At the satellite rotation rate of 100 RPM, 18.21 min are required to complete one full frame (see Figure 2.9).

The scanning mirror reflects the received radiation into a 16-in (40.5 cm) diameter telescope. A fiber optics bundle is used to couple the telescope to eight Visible band detectors (sensitive to the .54–.70 μm band). The fiber optics are configured such that each of eight Visible channels has a .25 × .21 mrad field of view, with the fields of view arranged in a linear array sweeping out the same scan line path. The field of view provides a nominal ground resolution of 900 m. The system, therefore, provides eight parallel west–east Visible data lines per west–east scan, covering the 8-km north–south band scanned by each step of the scanning mirror. In addition, germanium relay lenses are used to pass received radiation to two infrared detectors through an optical filter providing a 10.5–12.6 μm band pass. The resolution of the infrared sensor is nominally 8 km, corresponding to one north–south step of the scanning mirror.

The output from the eight Visible detectors and one (or an average) of the two infrared detectors is digitized on-board the satellite and transmitted in real time to earth. The quantitization of the infrared data is 8 bits and of the Visible, 6 bits. The infrared data are oversampled in the west–east direction,

FIGURE 2.9. One full frame from VISSR scanning system, showing full disc coverage.

so that the sampling resolution is 4 km in the west–east direction; this leads to the designation of "raw" infrared data as "8 × 4 km" resolution.

The real time data are received at Wallops Island, Virginia in bursts as the rotating mirror scans the earth. The data are formatted; calibration of the infrared is performed; grid, orbit, attitude, and various other items of information are added; and the data are retransmitted at a lower data rate back to the satellite for relay to various user stations. The data rate is selected so that the retransmission just occupies the interval between real time data bursts; that is, the retransmission is accomplished while the rotating mirror is viewing space between scans of the earth. These retransmitted data are known as *stretched* VISSR inasmuch as their time scale is stretched by virtue of the reduction in data rate.

Data from the VISSR Archive Tapes are available from SDSB. The requester need only specify the satellite (by name or simply as east or west) and the date and a copy will be made of all of the data archived for that date. Each day's data for each satellite occupy one tape (at 1600 bpi). If not otherwise specified by the user, copies will be written at 1600 bpi (9-track). Orders for tape copies that are not too extensive can generally be filled within 10 days. Service other than tape copying, such as data printouts and the like, are special orders and the cost and servicing time will vary with the complexity of the request.

The archived VISSR data cover a rectangular subarea of a complete frame, called the Archive Sector. The normal geographic coverage of the Archive Sector is about 88° latitude by 99° longitude for data prior to 14 November 1977 and 105° by 99° thereafter. This coverage is reduced when a satellite is operated in limited scan mode, as in the case of storm days, when data are collected only as far south as 5°S. Even on normal operating days, some variation in coverage will occur.

The Archive Sector is centered on the satellite subpoint in the east–west direction and usually will start at 50°N. The process by which this area is extracted involves the use of orbit and attitude information and the operational Earth-location software, the accuracy of which is about 10 km.

The infrared data are archived at a resolution of 8 km in both directions (every other sample along each scan is dropped). The infrared samples are 8-bit values (called *counts*), packed one per 8-bit byte. The Visible data are sampled and averaged from its original resolution of 1 km in both directions to 8 km. The Visible samples are 6-bit counts packed right-justified into 8-bit bytes.

The infrared data may be calibrated by the user in terms of temperature. The accuracy of these temperatures is limited to 2–4° Kelvin (K), generally nearer 2°K in the winter and 4°K in the summer. During the eclipse periods, 3 weeks on either side of the equinoxes, an additional 4°K error occurs due to thermal fluctuations on-board the satellite as it moves through the earth's shadow.

Each tape contains the data archived from one satellite's data for 1 day. Each Picture File nominally consists of 1373 records: a 320-byte Header Record, four 6720-byte records containing a 26,880-byte Benchmark Table, and 1368 Data Records, each containing the data from one scan or line of the

SMS–GOES picture. The maximum number of Data Records is 1368, the full archive image contains a maximum of 1368 lines. A sector limited in north–south extent would contain fewer lines. The length of a Data Record corresponding to a complete line is 1628 bytes. The length of Data Records of a sector limited in east–west extent will be less than 1628 bytes. The Header Record of each Picture File will include the number of Data Records in the Picture File and the length of each. The length of the Data Record is constant with the Picture File.

EXAMPLE 2. VHRR

The VHRR instrument is a two-channel (.6–.7 μm Visible and 10.5–12.5 μm infrared) scanning radiometer system currently aboard the NOAA series of polar-orbiting satellites. Two VHRR instruments are installed on each satellite. Each radiometer consists of two sensors (one Visible and one infrared) that view the Earth by means of a rotating mirror. The mirror is aligned to provide cross-track scanning of the earth. As the mirror rotates, one sweep of a sensor's field of view across the Earth (and pre- and post-Earth space views, etc.) is called a *scan,* or *scan line.* As a result of the design of the scanning system, the normal operating mode of the satellite calls for direct transmission to Earth (continuously in real time) of alternating infrared and Visible scan lines from both radiometers—infrared from one and Visible from the other. The instantaneous field of view of both sensors is about .6 mrad, resulting in a nominal resolution of 900 m at the satellite subpoint for both Visible and infrared data.

The VHRR data are received at 3 NOAA–NESS ground stations: Gilmore Creek, Alaska; Wallops Island, Virginia; and San Francisco, California. The amount of VHRR data received during one pass of the satellite over the ground station is limited to the acquisition range of the station—a maximum of 50–60° of latitude along the orbital track is possible. The VHRR data are transmitted in analog form from the satellite and received and 8-bit digitized at the NOAA ground station. These digitized data, along with calibration information, are written onto magnetic CCTs that are placed in the 90-day archive.

An example of VHRR imagery is shown on p. 148.

The VHRR tapes are 9-track, 1600 bpi, phase-encoded, CCTs. Each tape contains one or more VHRR data sets. Each VHRR data set consists of two files. The first file is the Header File and contains only the Header Record. The second file is the Data File and contains a variable number of Data Records. Each file is terminated by a single End-of-File Mark. Each data record contains one Visible or infrared scan line with date–time and calibration information included. Visible and infrared scan lines are never mixed within a file. The Header record and Data Record formats are illustrated in Tables 2.2 and 2.3, to indicate the level of complexity involved. In addition, digital image tape products from SEASAT–A are listed in Table 2.4.

ELECTRONICALLY SCANNED MICROWAVE RADIOMETER (ESMR)

The ESMR (Electronically Scanned Microwave Radiometer), a sensor carried by the NIMBUS series of satellites, is sensitive to radiation from 19.225 to 19.475 gigacycles per second (GHz) (1.55 cm center wavelength). The cross-track scan ranges from $-50°$ to $+50°$ from nadir in 78 scan steps every 4 sec. The altitude is 1100 km. The beam width is $1.4 \times 1.4°$ near the nadir; the beam width at the 50° scan extremes is 2.2° cross track × 1.4° along track. Hence the resolution is 25 × 25 km at nadir and 160 × 45 km at the

TABLE 2.2
Data Record Format

Up to 6900 data records are written in each Data File, using 5340 characters (8-bit bytes). (Odd parity.)

The first 10 characters are: time code ASC II; bddd hh mm ss.[a]

Next 14 characters—binary counts.

11. Space view average
12. Space view average (same quantity-duplicated)
13. Voltage wedge: 0 volts
14. Voltage wedge: -1 volt
15. Voltage wedge: -2 volts
16. Voltage wedge: -3 volts
17. Voltage wedge: -4 volts
18. Voltage wedge: -5 volts
19. Voltage wedge: -6 volts
20. Visible calibration target
21. IR calibration target
22. T1 (thermistor in IR cal. target)
23. T 2 (thermistor in IR cal. target)
24. T3 (thermistor in IR cal. target)
25. Engineering data
26. Engineering data
27. 5340. Data—binary counts

[a]ddd = day (Julian Day); hh = hour; mm = minute; ss = second; b = blank.

TABLE 2.3
VHRR Mag-Tape Header[a]

Dec	Hex	Contents	Format
0	0000	Station identification (2MSC)	A
1	0001	Station identification (LSC plus $20)	A
2	0002	Mag-tape ID ("I" or "V")/MT# (1 or 2)	A/B
3	0003	Specified ending scan number	B
4	0004	Resolution (always 0–½ × ½ mile)	B
5:14	0005:E	Picture identification (20 characters)	A
15	000F	Day (MSD)	A
16	0010	Day (2 LSD)	A
17	0011	Hours (2 digits)	A
18	0012	Minutes (2 digits)	A
19	0013	Seconds (2 digits)	A
20:22	0014:16	Orbit identification (5 characters plus $00)	A
25	0019	SIU rate errors count	B
26	001A	Time code flag (0 = actual, 1 = operator def.)	B
27	001B	Forward–reverse flag (0 = F, 1 = K)	B
30	001E	Actual ending scan number	B
31	001F	Missing lines (IR or vis) count	B
46	002E	Extraneous interrupts count	B
47	002F	Buffer overflows count	B
48	0030	Mag-tape rate errors count	B
49	0031	Mag-tape parity errors count	B

[a] A = ASCII (8-bit characters); B = binary number or code; all unspecified words are unused and contain binary zero ($0000).

TABLE 2.4
Seasat–A Data Products

Sensor	Tape Products (9-TRACK, 1600-BPI)
Scanning multichannel microwave radiometer (SMMR)	Latitude, longitude, and time-located geophysical data of surface wind speed (no direction), sea surface temperatures, integrated air column liquid water and water vapor, rain rate, atmospheric path length correction, and individual channel brightness temperatures.
Synthetic aperture radar (SAR)	A limited number of SAR images, approximately 26, each measuring 25 × 25 km, are available on CCTs from this system.
Visible and infrared radiometer (VIRR)	A limited number of CCTs containing Earth-located, time-oriented, ocean radiance values are available (IR temperatures and VIS brightness).

end of the scan. The area of overlap between successive orbits permits complete global coverage in 12 hr.

Brightness temperatures are measured at each scan position. The effective brightness temperature range is 138–290°K. The imagery is designed to observe rainfall areas and variations in atmospheric moisture over ocean, snow, and ice formations, and areas of high soil moisture (Wang *et al.*, 1980) heavy rainfall, and snow cover over land masses.

ESMR data are collected at the Rosman and Alaska data acquisition facilities and relayed to the Goddard Space Flight Center. The data can be obtained on CCTs. Figures 2.10–2.16 show examples of NIMBUS III High Resolution Infrared (HRIR) imagery.

Commercially Available Electronic Data Acquisition

Customized electronic data acquisition using airborne sensors is obtainable from a number of commercial organizations. The sensors include multispectral scanners, synthetic aperture radar, infrared radiometers, and others. Nonimage data can be collected simultaneously using magnetameters, scintillometers, and such. The comparison of side-looking airborne radar with Landsat imagery over the same terrain is shown in Figure 2.17.

Rather than attempt to list here the various firms offering these services, the reader is referred to the advertisements for aerial survey services often found in *Photogrammetric Engineering and Remote Sensing,* Journal of the American Society of Photogrammetry.

DIGITIZING FROM FILM

Sources of Film Data

SKYLAB DATA

The NASA Skylab Program consisted of one unmanned and three manned missions. The unmanned space vehicle was placed in orbit in May 1973. The manned missions to the space vehicle were Skylab 2, launched on 5 May 1973 and recovered on 22 June 1973; Skylab 3, in orbit from 28 July 1973 to 25 September 1973; and Skylab 4, launched on 16 November 1973 and re-

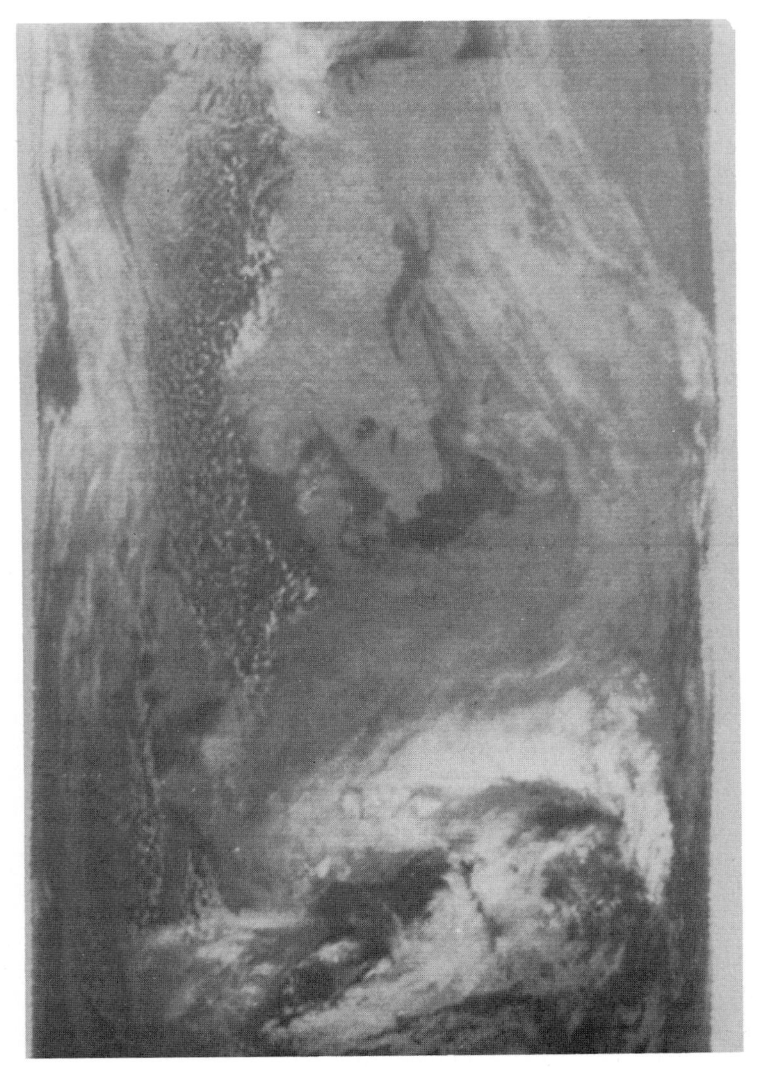

FIGURE 2.10. NIMBUS III HRIR nighttime orbit 3146, 4 December 1969, Western Europe.

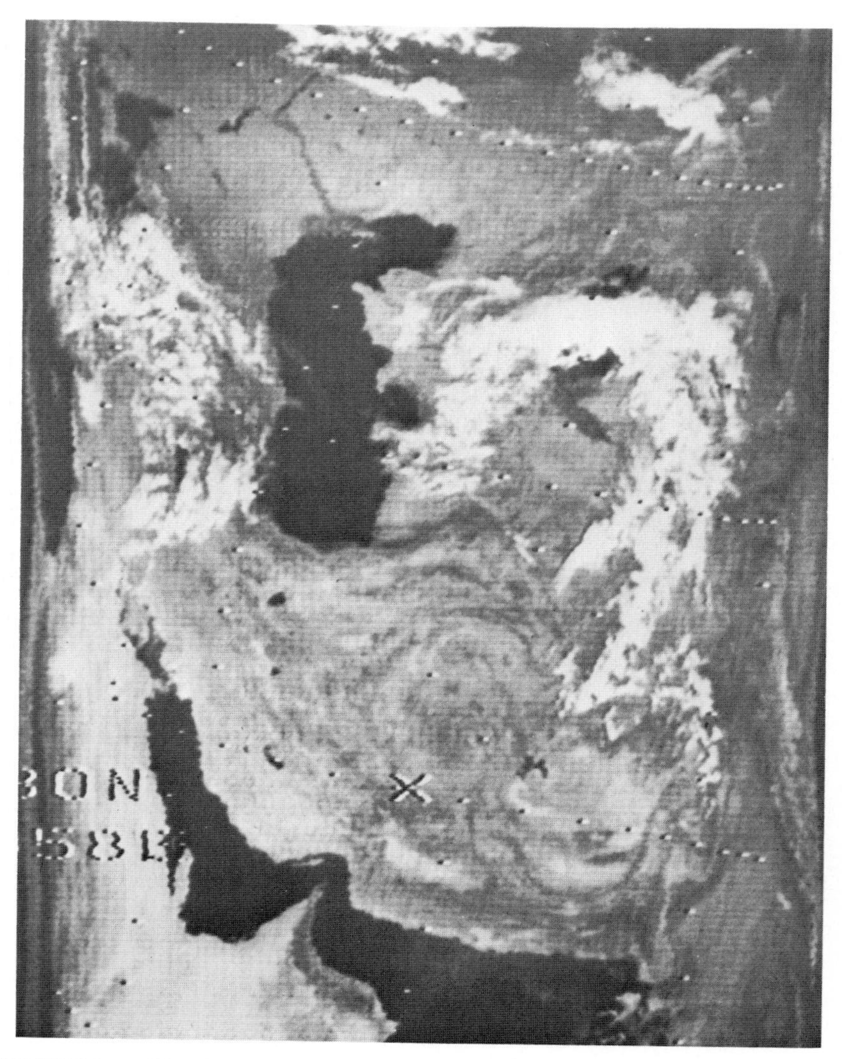

FIGURE 2.11. NIMBUS III HRIR daytime orbit 858, 17 June 1969, Volga River to Arabian Sea.

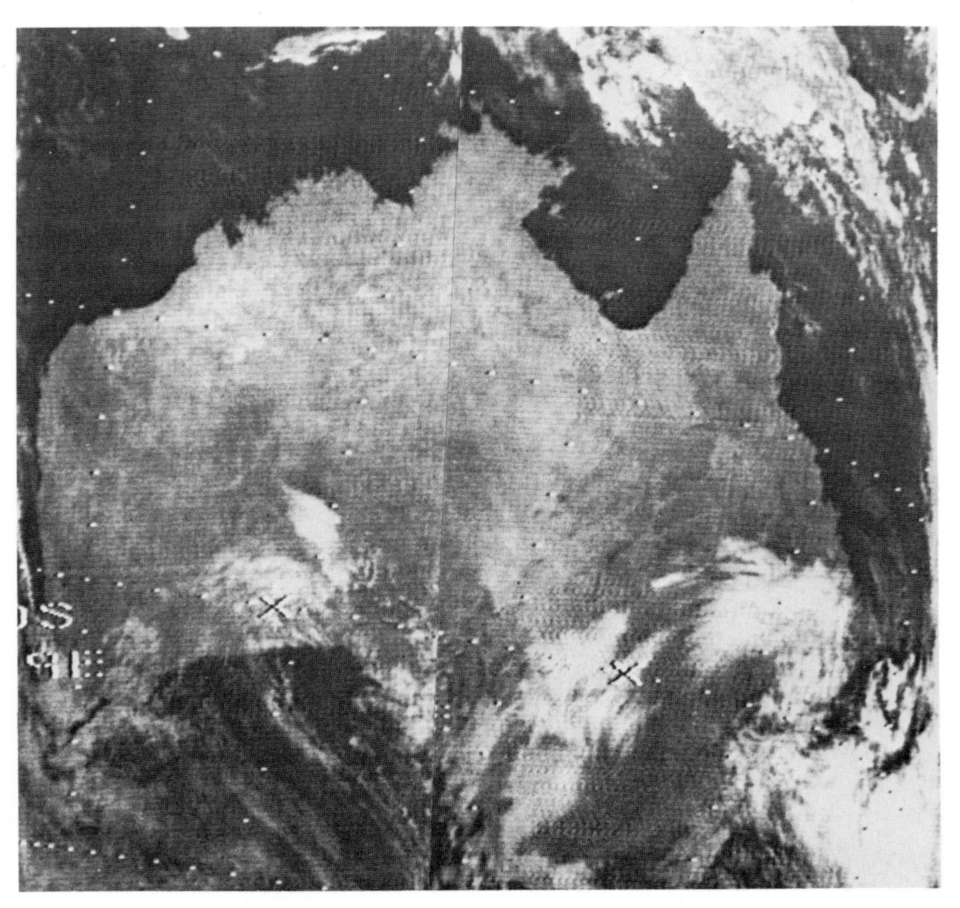

FIGURE 2.12. NIMBUS III HRIR daytime, June 1969.

covered on 8 February 1974. The spacecraft traveled in orbits 430 km (270 miles) above the Earth and acquired photography, imagery, and other data of selected areas between latitudes 50°N and 50°S. The data cover a number of scattered test sites selected to support Earth resources experiments. The photography, however, does not provide complete, cloudfree, systematic coverage of the Earth's surface. The Skylab Earth Resources Experiment Package consisted of three imaging remote-sensing systems:

FIGURE 2.13. NIMBUS III, direct readout IDCS, orbit 2295, 2 October 1969, Iberia and North Africa, from Apt, Stranraer, Scotland.

S190–A Multispectral Photographic Cameras. A six-camera array was designed to provide high-quality photography of a wide variety of phenomena on the Earth's surface. Each 152-mm (6-inch) focal length lens used 70-mm film. The films used were filtered black and white, color, and false-color infrared. The area covered by each image of this system is 160 × 160 km (100 × 100 miles.

S190–B Earth Terrain Camera. A single high-resolution Earth terrain camera was selected to provide high-resolution photography for scientific study. The 456-mm (18-in) focal length lens used 127-mm (5-in) film. Vari-

FIGURE 2.14. NIMBUS III HRIR daytime orbit 119, 23 April 1969, Southeast Asia

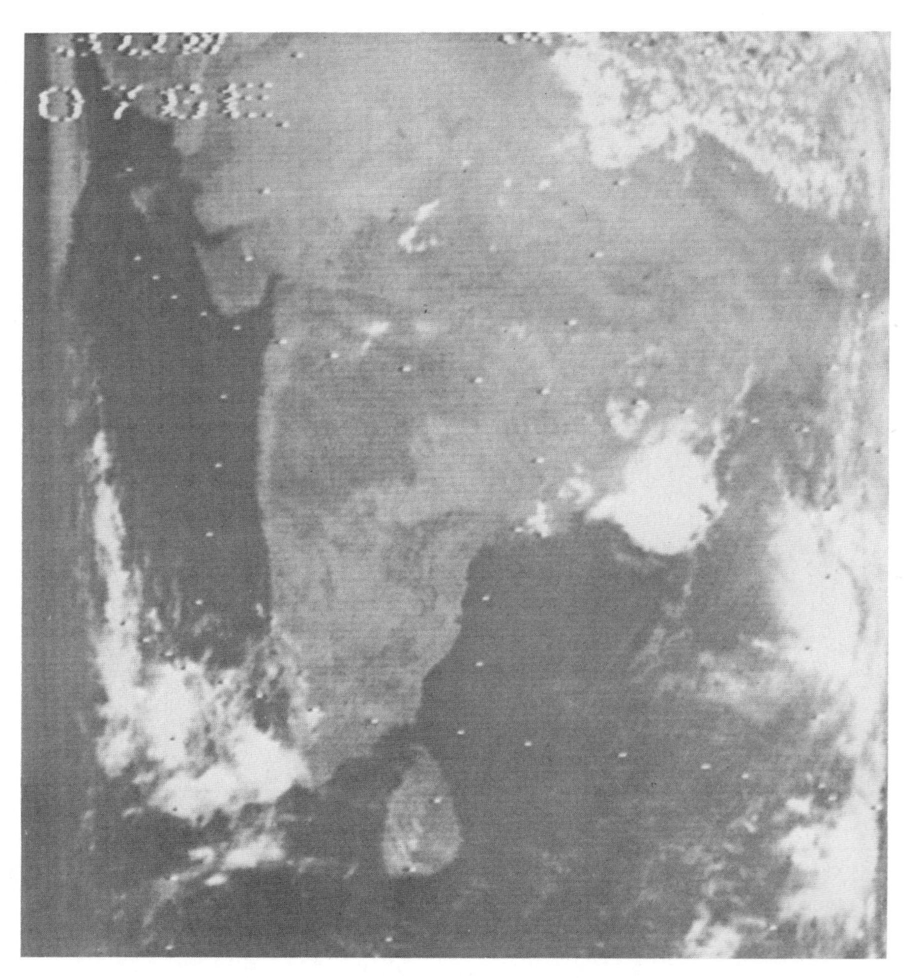

FIGURE 2.15. NIMBUS III HRIR daytime orbit 616, 30 May 1969, Indian subcontinent.

ous black and white, color, and false-color infrared films were used in the camera. The area covered by each frame of this system was 110 × 110 km (70 × 70 miles). An example of S190–B imagery is shown in Figure 2.18.

S192 Multispectral Scanner. Imaging sensor with 13 spectral bands, including one thermal infrared band, between .4 and 12.5 μm. Photographic data from the S190–A and S190–B experiments are available from the EROS Data Center. Data from the other systems are also available from the EROS

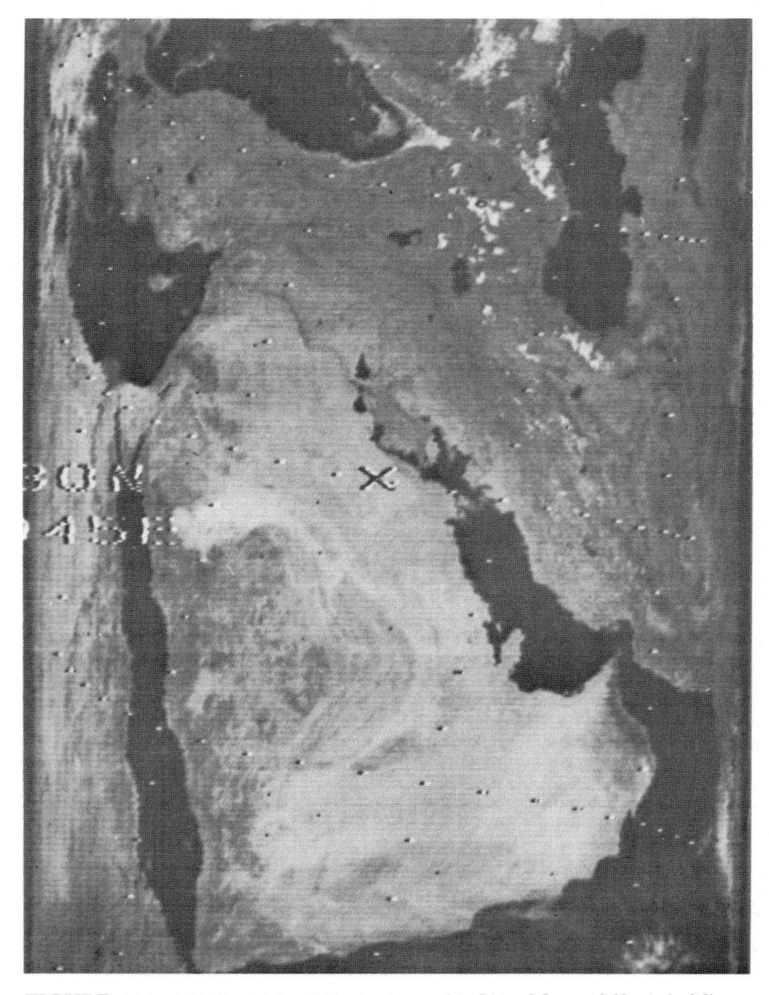

FIGURE 2.16. NIMBUS III HRIR daytime orbit 711, 6 June 1969, Asia Minor.

Data Center but must be individually requested from NASA by the Data Center for duplication and sale in response to an order.

NASA AERIAL PHOTOGRAPHY

NASA aerial photography is the product of aerial surveys carried out by the NASA Earth Resources Aircraft Program. The program is directed primarily

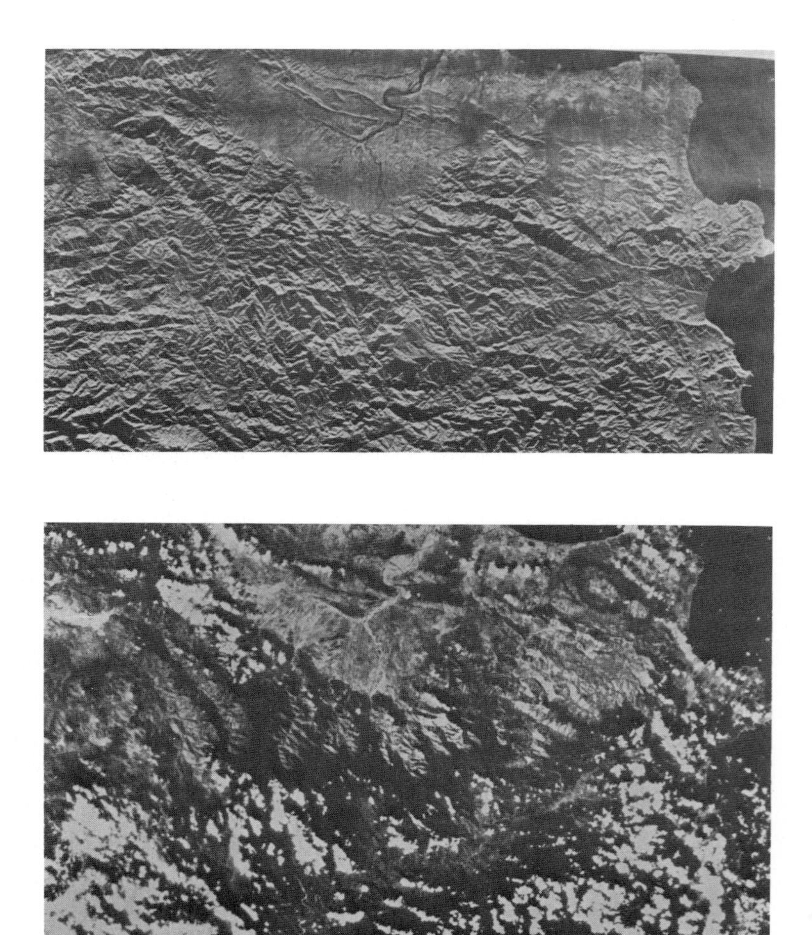

FIGURE 2.17. Comparison of side-looking airborne radar with Landsat imagery for Northern Luzon, Philippines.

at testing a variety of remote-sensing instruments and techniques in aerial flights, generally over certain preselected test sites within the continental United States.

Aerial photography is available in a wide variety of formats from flights at altitudes of a few thousand feet (1000 m) up to U–2 and RB–57F flights at altitudes above 18,000 m (60,000 ft). The high-altitude photography is generally available on a 23 × 23 cm (9 × 9 in) film format at approximate scales

FIGURE 2.18. An example of S190–B imagery.

INQUIRY FORM

GEOGRAPHIC COMPUTER SEARCH

U.S. DEPARTMENT OF THE INTERIOR
GEOLOGICAL SURVEY

DATE ______

NAME MR. / MS. ______ (FIRST) (INITIAL) (LAST) COMPUTER ACCOUNT NO. (IF KNOWN) ______

COMPANY ______ (IF BUSINESS ASSOCIATED) PHONE (Bus.) ______

ADDRESS ______ PHONE (Home) ______

CITY ______ STATE ______ ZIP ______ YOUR REF. NO. (P.O. GOV'T ACCT. OR OTHER) ______

TO INITIATE AN INQUIRY AND COMPUTER GEOSEARCH COMPLETE THE FOLLOWING

POINT SEARCH	POINT #1	POINT #2	POINT #3
LONG / LAT / Selected Point	Latitude ____ ° ____ ' N or S	____ ° ____ ' N or S	____ ° ____ ' N or S
	Longitude ____ ° ____ ' E or W	____ ° ____ ' E or W	____ ° ____ ' E or W

Landsat Only: (Worldwide Reference System)

Path ____	Path ____	Path ____	Path ____	Path ____
Row ____	Row ____	Row ____	Row ____	Row ____

Imagery with any coverage over the selected point will be included.

Return Completed Form To:

U.S. Geological Survey
EROS Data Center
Sioux Falls, SD 57198
FTS: 784-7151
Comm: 605/594-6511

For Additional Information or Assistance Please Contact One of the Following Offices of the National Cartographic Information Center (NCIC).

U.S. Geological Survey
National Cartographic Information Center
507 National Center
Reston, VA 22092
FTS: 928-6045
Comm: 703/860-6045

U.S. Geological Survey
Eastern Mapping Center
National Cartographic Information Center
536 National Center
Reston, VA 22092
FTS: 928-6336
Comm: 703/860-6336

U.S. Geological Survey
Mid-Continent Mapping Center

AREA RECTANGLE	AREA #1	AREA #2	AREA #3
LONG LONG LAT LAT	Lat. ____ ° ____ ' N or S to	Lat. ____ ° ____ ' N or S to	Lat. ____ ° ____ ' N or S to
	Lat. ____ ° ____ ' N or S	Lat. ____ ° ____ ' N or S	Lat. ____ ° ____ ' N or S
Imagery with any coverage over the selected area will be included.	Long. ____ ° ____ ' E or W to	Long. ____ ° ____ ' E or W to	Long. ____ ° ____ ' E or W to
	Long. ____ ° ____ ' E or W	Long. ____ ° ____ ' E or W	Long. ____ ° ____ ' E or W

If the above geographic coordinates cannot be supplied, please specify area by GEOGRAPHIC NAME AND LOCATION (include a map if possible.)

PREFERRED TYPE OF COVERAGE

	Black & White	Color or Color Infrared
☐ Landsat	☐	☐
☐ Skylab	☐	☐
☐ Nasa-Aircraft	☐	☐
☐ Aerial Mapping Photography (Minimum color available)		

PREFERRED TIME OF YEAR

Check maximum of three

- ☐ JAN-MAR
- ☐ APR-JUNE
- ☐ JULY-SEPT
- ☐ OCT-DEC
- ☐ All coverage
- ☐ Latest coverage
- ☐ SPECIFIC DATES ____

NOTE: Seasonal coverage normally applies only to Landsat coverage.

MINIMUM QUALITY RATING ACCEPTABLE

☐	☐	☐	☐
0-2 (VERY POOR)	3-4 (POOR)	5-6 (FAIR)	7-9 (GOOD)

MAXIMUM CLOUD COVER ACCEPTABLE

☐ 10% ☐ 30% ☐ 50% ☐ 70% ☐ 90%

NOTE: Classification of percent of cloud cover is subjective and is relative to the amount of clouds appearing on the imagery and not to their location.

APPLICATION AND INTENDED USE ____

FORM: 9-1936
(April 1978)

National Cartographic
Information Center
1400 Independence Road
Rolla, MO 65401
FTS: 276-9107
Comm: 314/364-3680

U.S. Geological Survey
Rocky Mountain Mapping
Center
National Cartographic
Information Center
Stop 504, Denver Federal
Center
Denver, CO 80225
FTS: 234-2326
Comm: 303/234-2326

U.S. Geological Survey
Western Mapping Center
National Cartographic
Information Center
345 Middlefield Road
Menlo Park, CA 94025
FTS: 467-2426
Comm: 415/323-8111

U.S. Geological Survey
National Cartographic
Information Center
National Space Technology
Laboratories
NSTL Station, MS 39529
FTS: 494-3541
Comm: 601/688-3544

PLEASE CONTACT THE NEAREST NCIC OFFICE FOR INFORMATION CONCERNING THE AVAILABILITY OF CARTOGRAPHIC PRODUCTS OTHER THAN IMAGERY.

FIGURE 2.19. The form used to request a geographic search by the EROS Data Center.

HOW TO REQUEST A GEOGRAPHIC SEARCH

This form is used to request a computer search for imagery over a point or area of interest.

Data from this inquiry sheet will be used to initiate a computer Geosearch. The results will be returned on a computer listing along with a decoding sheet, from which imagery can be selected and ordered.

Complete the form as follows:

A. Enter your NAME, ADDRESS, and ZIP CODE clearly. If you have had previous contact with that facility, include your COMPUTER ACCOUNT number. Enter a PHONE number where you can be reached during business hours.

B. Complete the required information for either the POINT SEARCH, or AREA RECTANGLE inquiry, which includes the geographic LATITUDE and LONGITUDE coordinates. If coordinates are not available, please supply the GEOGRAPHIC NAME AND LOCATION or a map with the area of interest identified. It is beneficial that you minimize your area of interest, thereby allowing for a faster and more critical retrieval of information.

C. Complete all other information.

D. Complete the APPLICATION AND INTENDED USE portion of the inquiry. e.g. Will it be used for identifying buildings or will it be framed and placed on a wall. This information will assist our technicians in determining whether the products available will satisfy your requirements.

E. Return completed form to the EROS DATA CENTER.

NOTE: If an inquiry is made for Landsat Data, and the Worldwide Reference of PATH and ROW numbers are available, please insert them in the appropriate locations. Otherwise, geographic coordinates will suffice.

FIGURE 2.19 *continued.*

of 1:120,000 and 1:60,000. In general, each high-altitude frame of 23 cm (9 in) film format photography at 1:120,000 scale shows an area approximately 27 km (17 miles) on a side.

AERIAL MAPPING PHOTOGRAPHY

Aerial photography during the past 25 years was acquired by the U.S. Geological Survey and other federal agencies for mapping of the United States. The photography is black and white and has less than 5% cloud cover.

Depending on the planned use of the photographs, the aerial-survey altitude ranged from 600 m (2000 ft) to 12,000 m (40,000 ft). The basic film format is 23 × 23 cm (9 × 9 in) and shows areas from 4.8 to 14.4 km (3–9 miles) on a side, depending on the scale of the photograph.

Because of the large number of aerial photographs needed to show any specific region on the ground, the photographs have been indexed by mounting series of consecutive and adjacent overlapping photographs to create mosaics of photographs of specified areas. These aerial photographic mosaics are referred to as *photo indexes* and allow for rapid identification of photographic coverage of a specific area. Some 50,000 photo indexes are available at the EROS Data Center. When ordering aerial photography from the EROS Data Center, it is necessary first to order a photo index of the area of interest to determine the specific aerial photography needed. The form used to request a geographic search by the EROS Data Center is reproduced as Figure 2.19.

GEMINI–APOLLO PHOTOGRAPHY

Photographic coverage acquired by the Gemini and Apollo missions over limited areas of the Earth is available from the EROS Data Center. The photographs from the Gemini missions were acquired by hand-held 70-mm cameras. Table 2.5 lists Gemini and Apollo photographic data.

A special photographic experiment, designated S–065, was conducted on Apollo 9 in March 1969. This experiment used four 70-mm cameras to acquire three types of filtered black and white and one type of false-color infrared photographs. Each picture shows an area approximately 160 × 160 km (100 × 100 miles).

Digitizing Equipment and Services

The list of image processing hardware vendors on page 52 includes companies that make and sell scanning film digitizers. Some of these vendors also

TABLE 2.5
Gemini–Apollo Photographic Data[a]

Mission	Launch	Recovery	Photographic coverage
Gemini III	23 March 1965	23 March 1965	25 color photographs of the Earth.
Gemini IV	3 June 1965	6 June 1965	219 color photographs of the Earth, including extensive coverage of the Southwestern United States.
Gemini V	21 August 1965	29 August 1965	250 color photographs of the Earth.
Gemini VI	4 December 1965	16 December 1965	192 black and white and color photographs, primarily of Africa.
Gemini VII	4 December 1965	18 December 1965	429 varied black and white and color photographs.
Gemini VIII	16 March 1966	16 March 1966	19 black and white and color photographs, mainly of the Agena launch missile.
Gemini IX	3 June 1966	6 June 1966	362 black and white and color photographs of the Southwestern United States, west-central South America, and Africa.
Gemini X	18 July 1966	21 July 1966	371 varied black and white and color photographs, which include some coverage of China, Central America, and the Straits of Gibralter.
Gemini XI	12 September 1966	15 September 1966	238 black and white and color photographs of the Earth including the Middle East, North Africa, and Australia.
Gemini XII	11 November 1966	15 November 1966	403 assorted black and white and color photographs, covering many areas of the globe.
Apollo 6	4 April 1968	Unmanned	370 varied Earth-looking color photographs taken by a stationary camera.
Apollo 9	3 March 1969	13 March 1969	912 varied Earth-looking color photographs of selected test sites.

[a] Approximately 1000 of the above photographs have been evaluated as acceptable Earth-looking images and are presently available in the geographic inquiry system.

offer film digitizing services. A photograph of the Itek film scanner is shown in Figure 2.20. Chapter 6 includes a tabular list of this scanner's performance specifications.

Typically, a laser beam is directed along the axis of a lathelike rotating drum mechanical assembly. A 45° angled mirror redirects the beam at 90° to the axis so that it shines onto the inner surface of the rotating drum. A window in the drum allows the beam to pass out from the drum for some fraction of each drum revolution. As the beam exits the drum, it impinges onto a light detector producing an electrical signal. A clear window produces a square wave output signal from the detector as the drum rotates. By inserting a photo transparency over this drum window, the laser beam is caused to scan a line

FIGURE 2.20. A photograph of the Itek film scanner.

across the photo; thus the signal from the detector is modulated by the variation in photographic density from point to point along this scan line. Opaque points let little light reach the detector, whereas clear points on the photo diminish the light transmission very little.

Now, if the drum is allowed to move along the axis very slowly, a raster pattern will be scanned over the entire window, one line per drum rotation. The amount of movement along the axis per drum rotation is chosen to match the beam diameter at the window; the beam is brought to a focus using a lens assembly with the focal point at the film surface. The analog signal from the detector element is amplified and sampled to break it into pixels, and each sample is quantized by an analog to digital converter. The results are then stored on magnetic tape.

REFERENCES

Holmes, Q. A., Wensch, D. R., & Vincent, R. K. Optimum thermal bands for mapping general rock type and temperature from space. *Remote Sensing of Environment,* 1980, *9*(3), 247–263.

Hovis, W. A., Clark, D. K., Anderson, F., Austin, R. W., Wilson, W. H., Baker, E. T., Ball, D., Gordon, H. R., Mueller, J. L., El-Sayad, S. Z., Sturm, B., Wrigley, R. C., & Yentsch, C. S. Coastal zone color scanner: System description and initial imagery. *Science,* 1980, *210,* 60–63.

Traganza, E. D., Nestor, D. A., & McDonald, A. K. Satellite observations of nutrient upwelling off the coast of California. *Journal of Geophysical Research,* 1980, *85*(c7), 4101–4106.

Wang, J. R., Shiue, J. C., & McMurtrey, J. E. Microwave remote sensing of soil moisture content over bare and vegetated fields. *Geophysical Research Letters,* 1980, *7*(10), 801–904.

Computer Processing

EQUIPMENT

Processing Equipment

There are two essential hardware ingredients in a digital image processing system: a computer and a display device. The computer may be general purpose or specifically designed for image processing (Johnson, 1975). Its size can range from micro to the enormous, advanced computers that execute hundreds of millions of instructions per second (Potter, 1978), depending on the volume of image processing work to be performed. Tape handling capabilities are particularly important in image processing activities because of the large volumes of data employed (Danielsson, 1980). Table 3.1 is a list of image processing hardware systems vendors.

Digital image processing software systems are available from many sources. The government clearinghouse for public domain image processing software is COSMIC, 112 Barrow Hall, Athens, Georgia, 30601. The COSMIC repository includes VICAR, SMIPS, LARSYS, ASTEP, and LACIE, to name a few. Another software system of interest is available from Professor Robert Haralick, Virginia Polytechnic Institute and State University; the system is named KANDIDATS (Haralick & Minden, 1978).

TABLE 3.1
Image Processing Hardware Systems

Aydin Controls
414 Commerce Drive
Fort Washington, Pennsylvania 19034

Broomall Industries, Inc.
682 Parkway
Broomall, Pennsylvania 19008
(GP–100 Graphics Processing System)
(VISICON)

Comtal Corporation
169 Halstead
Pasadena, California 91107
Contact: John Tahl
(Comtal Series 9)

Control Data Corporation
P.O. Box 1249
Minneapolis, Minnesota 55440

DICOMED Corporation
9700 Newton Avenue South
Minneapolis, Minnesota 55431
(Image Data Processing Peripherals)

DIGICOM, Inc.
Paramount Building
Chelmsford, Massachusetts 01863
(DIGICODER)

DeAnza Systems, Inc.
3444 DeLaCruz Boulevard
Santa Clara, California 95050

ESL, Incorporated
495 Java Drive
Sunnyvale, California 94086
(IDIMS)

Grinnel Systems
2986 Scott Boulevard
Santa Clara, California 95050

I/O Metrics Corporation
1050 Stewart Drive
Sunnyvale, California 94086
(LASCO System)

Interpretation Systems, Incorporated
P.O. Box 1007
Lawrence, Kansas 66044
(ISI System 470 and SAND)

General Electric Company
5030 Herzel Place
Beltsville, Maryland 20705
(Image 100)

LEXIDATA Corporation
215 Middlesex Turnpike
Burlington, Massachusetts 10803
(System 6400)
(Model 200-D & 400-D)

Lockheed Electronics Company
Remote Sensing Applications Lab
1830 Space Park Drive
Houston, Texas 77058

Bendix Aerospace Systems Division
Ann Arbor, Michigan 48107
(M-DAS)

Ford Aerospace & Communications Corporation
P.O. Box 58487
Houston, Texas 77058
(M-DAX)

MacDonald Deltwiler & Associates
Ste. 109, 2182 W. 12th Avenue
Vancouver, B.C., Canada V6K 2N4

Optronics International, Inc.
7 Stuart Road
Chelmsford, Massachusetts 01824

Ramtek
585 North Mary Avenue
Sunnyvale, California 94086

STARAN
Goodyear Aerospace Corporation
Akron, Ohio 44315

Sci-Tex North American
1450 Broadway
New York, New York 10018

Spacial Data Systems, Inc.
Box 494
508 S. Fairview Avenue
Goleta, California 93017
(Data Image Station)
(Data Eye Com)
(Data Color)
(Computer Eye)

International Imaging Systems
650 North Mary Avenue
Sunnyvale, California 94086
(System 101)

Figures 3.1 and 3.2 show differing approaches to the design of an image processing system. Figure 3.1 shows the installation view of the General Electric IMAGE–100 System. This is a well-packaged, well-documented stand-alone image processing system designed primarily for high productivity on multispectral classification applications. Some algorithms are implemented in special purpose hardware. General computation is done in a PDP–11/40. Figure 3.2 shows a much earlier configuration. It is the IDECS (Haralick & Currier, 1977) image processing system shown in use at the Engineer Topographic Laboratories, Fort Belvoir, Virginia. This system was designed primarily for algorithm and special purpose device research where ease of reconfiguration is important.

Display

Adequate image display capabilities are essential for effective image processing. Three types of display devices are in common use. The conventional computer line printer is the simplest, least expensive and most readily available display device. Typically, the computer is programmed to represent an image on a line printer in the form of a shade print. Each pixel is coded as a

FIGURE 3.1. Installation view of the General Electric IMAGE–100 System.

set of overstruck characters; white or light shaded pixels are represented by a blank or a dash or another light shaded character. Dark pixels are shown on a shade print as an *X* overstruck by an *H* overstruck by an (#). Together these make a fairly dark spot on the printer paper. Intermediate gray shades are shown by other character combinations. Figure 3.3 shows a sample shade print at full scale. The image-like nature of the shade print is best discerned by looking at it from a distance so that individual pixels are not seen. The shade print is also quite convenient in circumstances that call for counting individual pixels or for marking specific features for reference with a pencil. Figure 3.4 shows another way to present shade print information. Here eight shade print sections have been pieced together to cover a large wall. This mosaic shade print is then photographed and printed as a negative at a reduced scale.

Another type of display device finding widespread use is the CRT. In its basic form, it is identical to the ordinary television picture tube, although special high-resolution CRTs have been developed specifically for image processing applications. In all cases, special electronics must be incorporated

FIGURE 3.2. The IDECS image processing system, shown in use at the Engineer Topographic Laboratories, Fort Belvoir, Virginia.

with solid state refresh memories to allow the digital imagery to be displayed. This approach is particularly useful in interactive image processing environments. Some drawbacks include the CRT's small format (about 500 × 500 pixels), its degraded image quality as compared with photographs, and a lack of hard copy for record keeping.

The third type of display device is the digital film recorder. Here a raster scanning light beam is modulated by the digital pixel gray-level values and is caused to expose photographic film. Figure 3.5 shows the loading of film into such a recorder. After exposure, the film is either developed automatically and internally with wet chemicals or Polaroid process or it is removed for conventional development in a darkroom. These recorders are comparatively expensive and require substantial maintenance. But they do offer high image quality, large formats, and archivable hard copy.

PREPROCESSING

Before the various digital enhancement or analysis processes are applied to a new digital image data set, there are several preprocessing operations one may choose to perform. In general, these either facilitate later operations or help the analyst characterize the behavior of the data. Six of these are described in this section, although many others are in use at some facilities. Among the preprocessing operations not addressed here are ortho-processing, which removes the effects of an oblique look-angle, and temporal normalization, which adjusts the image to standardize the apparent sun angle. The operations described here are in much more general use.

Reformat

As described earlier, newly arriving digital image data sets come in a bewildering variety of formats. A facility engaged in processing a number of these will find it convenient to reformat this data by computer into a conventional format adopted as the standard for that facility. Even if only one type of incoming format is encountered, say Landsat, experience has shown that reformatting the tapes into a simpler data arrangement is beneficial. The software used later to process this data is in turn made simpler, and the resulting saving in computer time generally makes up the cost of the reformatting by a good margin.

One attractive format that is useful as a facility standard is for each image

```
SUBIMAGE SHADE PRINT:

STARTING AT PIXEL (1950,1951)

     250 ROWS WERE PROCESSED
     250 COLUMNS WERE PROCESSED

  2081      2091      2101      2111      2121      2131      2141
     |+++++++++|+++++++++|+++++++++|+++++++++|+++++++++|+++++++++|+
RCW
1950 88.---.@@@≡++/@≡///+++/â≠++//+â8++≠≡+..≡≡+...-.+//+./≡88€/.≡8
1951 ≡/≡.-+C@≡/≡+..+≡C@@8/+///≡≡/+≡+.+≡++@≡.+.+≡≡+--..//+++./888O.O
1952 88≡+≡€€@≡≡@≡+./≡8888/+.++/@≡.//./â≡+≡@/+.+≡≡+..+./≡/++≡≡@@@888
1953 C€8CC8800≠≡/=--≠€CC≠/=/≡/≡0≡=≡C//€C=≠8≡..=≡≡==--/≠0≠00≠≠¬/≡≠08
1954 000000000≡/≡≡===/≡≡≡≡//===00//≡/=≡≡//@≡===//////≡≡≡≡@0≡≡≡≡≡≡88
1955 @@8888@@@@≡+8≠.-./≡≡≡+.+/++≠8≡/≡≡+≠≠+/≡≡≡@€≡++./≡≡≠≠≡≡@≠≡≠//≡@@
1956 8@@8@8888≡+@@.+./≡≡≡/.-+++/8≠+≠@+/88≠-.888@+/+/≠≡≠≡≡≡@≠@≠≠≡≡≠8.
1957 888@@8888@≡00≡≡..+≡+//..../C++8C--@€C--C8€C+++/≡++≡@€@@80≡0≡/8.
1958 8€88@€88@≡≡8@≡+..+≡@≡//≡â@8/--@@--≡8€--/@8≡//+//≡≡@@≡≡@≡≡≡/≡≡8
1959 880≡/≡==.-≡88≠//≡≡.- =€€CCC/--8€.-/€€.-/88≠////≠0≠00880≡≠0//≠8
1960 @≡/=//..≡@0≡---088/  -â8€888€=-≡C/--80=--08â≡/==/≡//0@≡0@@@≡/â8
1961 +/≠≡+- -≠88+  -≠88/  -≡8€@@â--+@≠--≡8≠--≡8≡≡/++++//≡/≡≠≠≡≠≡+≡8
1962 +@8≠- --888≡- -+88≠---+€8888≠-+8≠--+8€.-+88@//++/≡≡≡≡@8@≡+.--.
1963  -//0+ --+0@-   .00≡≡---.888888-.â€/-+88--+88â/++/≡0≡≡+..-----
1964  -+≡/.  --≡≡.   -≡≡@@.---€€€888.-≡€≡-.8€+-.888≡.++≡88@.-----../
1965   -/≠≡-  -≡≠/-   ≡≡80=---≠88088=-≠€≡--08≡.0888≠=..0880=../C8€8
1966   -≡@≡.  -==-.  -//00≡---≡€€888/-@8C≡≡8€@≡≡€8C/.../888≡=.≡8888
1967   -/@@+  --≡//-  --≠@≠--  .888888≠/≡≡â88≠@≡/@8€@+-+/888≠..+88≡/
1968   -./≠≡-   --+-----≠88+---€8/≡@≠@≡/≡@@≠≠≠≡≡8888+---≡88â+...+.+
1969 .  --@≡-    -..-----0≡+≡â+.€8++0â≡+/0@8€@@≡+≡888+-.-≡88@+.++++
1970 .- --/≡/-- -≡≡+≡@+++@≡+≡€≡.â8++88≡≡≡≡≡≡≡@@≡≡//888≡-../888/..+≡≡
1971 =-  -≠0≠/≠/.C≠=/0≡≡.≡≠.=CC.≠0=.08≠¬≡≠0≠C0≡/≠≠€88C.../888≡-.=≠0
1972 =/=..≡@=.â≡=≡â/.âââ==≡≡=C8=/@/.≡8@≡@≡≡≡€888€8888€=--=888@=..=.
1973 88@≡-/@/+≡≡++@≡.≡≠≠..≠≡.≠8/+@≠.+@≠88≡88888888888888/--.≠≠++-----
1974 888≡++8≡./@+.≠≠-+€8/.≠8./8≡-≡@++€888888888888888888+.-//-.---+/
1975 88888++@/-/@/.≡C.+8≡.≡C/≡@≡+≡8/0888888888888888888/.-.≡@@≡8888
1976 8@888≡.@@./€≡.≡â//8@.≡≡≡≡8â./â//8888888888888888888≡≡++++≡+@888
1977 88888€0.C8/.00=≡80≠88..CCC88=/80/8888888888888888888=-.--.=-.≠088
1978 @€888C/≡8@=08088€888≡=88=≡≡=≡/-.≡8888888888888888≡=----------≡8â≡
1979 ≡≡.+≡€€@€88888€€88@8≠//----.≠8+--≠8€88888888888≠.-------
1980 ----+8888888888@/≡+++-------..++//@≡@8€8888888888@.-----
1981 .----.88888≡/+++...+/+-------08@08888888€8888888+-
1982 -.++/≡888€+++.-.+++/++.--++≡88888€€€€€8€€88€€888/
1983 C8888€8888C...-..///.==≡C€€€88888€888888888888888/
1984 8888888888/.===.=..--≡€€€88888888€€8CC880888888888≡
1985 88888€€8≠≡+..-------+.≠88€8888888€€€888888888888888.
1986 888≠///+++....//≡≠≠8€€888888888€€€€€888@88808888≡+-
1987 88/+....+...≡888€€888888€88888888888@88888888€€8≡+...----- -+/
1988 ≡8≡+-..+≡≡@€€€€€888888â≡/≡â@8888€888888888888888≡//+./++++/+-+88
1989 80≠≠≠88€88888888C≡≡./=/=-.--≡888€€€€C€888€€88C≠/==//≡/.../==≡€8
1990 8≡08888888C≡≡0≡---..-.=.=≡≡/=//..≡8€C8888C≡...--/=///.=//C88888
1991 8@@@≡//≡-.//.--.+///+≡+.+--------≡€€€8@+-..-..--//≡@@8€88888€≠â8
1992 8≠≠+---+.//-../+...---../.-≡≠≡/≡ââ€8≠/+.+.-..--@88888≠≡≠≠+/≠88
1993 88@@/-+≡888/-----.-...+C≡+-+≡//CC€8€//+-.+--/+.≡80++≡@â8888888
1994 8@≡8@+..///.--+++≡/+++.//++≡â≡≡88€≡++//+/≡+.+/./88888888888888
1995 CC000=.≡000/==-===.-.=/≠C≠≡≠888888≡≡/=≡≠/≠≡.=//≡88888888888888
1996 0880@≡≡@≡≡≡+...=≡≡≡@≡≡≡+++=≡888888ââ≡.≡≡+≡≡==+==08888888888888
1997 ≠@88@≠++≡≡≡/≡≡/./.-.//≠≠/+-.≠88888â≠≡+.+.+/.+88888€8888888888888
1998 @≠≠≠@8@≡≠≡/..+/≡≠≠≠≠≠≡+.-.+â888888€€€≡+@8@888€8888888888888€€888
1999 8@@8@8@++≡0@8@≡/C88888≡+≡@8€88888€€€€88888888€€€88€€€88888888888
2000 88@88888≡≡/+.++@€€€€€888€€@@8@≡888888€€888888€88€€888€888888888888
2001 88888€8≠///./C€€88888≠/≡≠≡≡CCC≡008CC≠€88€€€€€€€€888888888888888
2002 888888C@@888888€88888@â≡++=+-.---..=+0888888888888888888888888888
```

FIGURE 3.3. A sample shade print at full scale. The image-like nature of the shade print is best discerned by looking at it from a distance so that individual pixels are not seen.

2151 2161 2171 2181 2191

++++++++|+++++++++|+++++++++|+++++++++|+++++++++|+++++++++|++++

FIGURE 3.4. Eight shade print sections have been pieced together to cover a large wall. This mosaic shade print is then photographed and printed as a negative at a reduced scale.

FIGURE 3.5. Film being loaded into a digital film recorder. A raster scanning light beam is modulated by the digital pixel gray level values and is caused to expose photographic film.

or each band of a multispectral image to be recorded as a file followed by an end-of-file mark. Each line of that image is recorded as a record in that file. The first record of the file is a header record. There is 1 byte per pixel. This format is simple and matches the format requirements for input to several image display devices. It also satisfies the NATO standard for the exchange of digital image tapes among member nations. No program listing is given here since this depends on characteristics specific to each computer installation.

Editing

A word in passing must be said about editing. In some circles this operation is termed *windowing*. It consists of a rather simple utility program or subroutine that accepts as input a digital image and the coordinates of the upper-left and lower-right corner of a rectangle within the format. The output is the digital image contained by that rectangle. The upper-left corner of the rectangle in the output has coordinates (1,1). The pixels outside the rectangle are simply discarded. This simple operation deserves special mention to emphasize that digital images contain a lot of data. Without proper editing early in the processing sequence, a great deal of processing effort and cost can be expended wastefully.

Gridding

To make distance measuring easier, it is often convenient to superimpose a rectilinear grid on an image. Gridding schemes can be simple or elaborate. One simple method is to insert a white (or black) pixel after every tenth pixel in every line and insert a white line after every tenth line in the format. An elaboration on this plan is to make every tenth inserted pixel and every tenth inserted line of a contrasting gray level with the other inserted grid elements.

Where scale preservation or format size is of greater importance than information preservation, the grid elements can replace pixels rather than being inserted between pixels. Another algorithm for gridding, which is used primarily in the replacement scheme rather than in the insertion style, is to decide, on a grid element by grid element basis, whether the grid element is white or black, depending on whether the replaced pixel is dark or light. This method has the advantage of keeping the grid visible everywhere, even if the local image content varies over wide extremes. An example of gridding a digital image is shown in Figure 3.6, which is based on a window of a Landsat frame containing an image of Honey Lake, California.

Annotating

There are three basic forms of digital image annotation. These may be termed *designating, captioning,* and *calibrating*. Designating consists of superimposing arrows, circles, rectangles, crosses, or other graphic entities on the image itself. In this way, points of features of special interest can be identified for reference. The elements of the graphic character are simply written over the gray value on the pixels under them, and hence some image information is lost. The most common use of designating in digital image processing is the insertion of latitude and longitude lines and numbers on images derived from meteorological satellites.

Captioning adds binary lines at the bottom of a digital image. Programs for this operation generally accept descriptive alphanumerical information up to, say, 120 characters. The program converts these characters to their matrix representation of black and white dots. This collection of 7×9 matrix items

FIGURE 3.6. An example of gridding a digital image that is based on a window of a Landsat frame containing an image of Honey Lake, California.

are strung together to compose nine image lines appended to the digital image itself.

Calibrating annotation is added primarily to check the fidelity of the image recorder employed. A known pattern is attached to the image to assure the analyst that the image display device is accurately presenting the digital information. These patterns commonly include step wedges, resolution targets, and starburst angular resolution targets. For color imagery, full Visible spectra are added to the image; as with captions, these calibration patterns are generally added below or beside the image format.

Histogram

The histogram (Pratt, 1978; Rosenfeld & Kak 1976) of an image is a table or graph showing the number of occurrences of each gray level. Examples of histograms and their uses are shown on pages 117 and 142.

Entropy

One quantity frequently calculated during preprocessing is the entropy of the picture, which is a measure of the picture's information content. A concept derived from information theory, *entropy* quantifies the degree of organization or disorganization in an image. The entropy is also the number of bits required to encode the pixels of an image without loss of information (Pratt 1978; Rosenfeld & Kak, 1976).[1] Image manipulations that increase the entropy enhance the image. Therefore, entropy is used as one measure of the success of an image enhancement operation.

To compute the entropy of an image, the histogram is employed. For a picture consisting of M pixels ranging over N possible gray levels, let the histogram function be $h(n)$, $n = 0,1,2,\ldots, N-1$. Then define $p(n) = h(n)/M$. Then $p(n)$ is the probability of a pixel having gray level n.

The entropy is defined as

$$H = - \sum_{n=0}^{N-1} p(n) \log_2 p(n)$$

Perhaps this is best illustrated by example. Consider a digital image consisting of 64 pixels ranging over eight possible gray levels:

n	$h(n)$	$p(n)$	$\log_2 p(n)$	$p(n) \log_2 p(n)$
0	2	.031	−5.000	−.155
1	0	.000		.000
2	6	.094	−3.411	−.321
3	18	.281	−1.831	−.515
4	20	.313	−1.676	−.525
5	10	.156	−2.681	−.418
6	5	.078	−3.680	−.287
7	3	.047	−4.411	−.207
Total	64	1.000		−2.428

Hence $H = 2.428$. On the other hand, if $h(n)$ were equal to 8 for all values of n, then H would be equal to 3.000.

[1]An example of this coding is given on page 115; run length codes and other coding schemes can reduce the bits per pixel below H.

REFERENCES

Danielsson, P. E. *The time-shared bus—A key to efficient image processing.* Paper presented at the 5th International Conference on Pattern Recognition, IEEE, Miami Beach, Florida, December 1980.

Haralick, R. M., & Currier, R. Image discrimination enhancement combination system (IDECS). *Computer Graphics Image Processing,* 1977, *6,* 371–381.

Haralick, R. M., & Minden, G. KANDIDATS: An interactive image processing system. *Computer Graphics Image Processing,* 1978, *8,* 1–16.

Johnson, R. G. M–Das—System for multispectral data analysis. *Proceedings of the NASA Earth Resources Symposium, Houston, Texas* 1975, *1-B,* 1323–1350.

Potter, J. L. The STARAN architecture and its application to image processing and pattern recognition altorithms. *Proceedings of the National Computer Conference. 47,* 1978, 1041–1047.

Pratt, W. K. *Digital image processing.* New York: Wiley, 1978.

Rosenfeld, A., & Kak, A. C. *Digital picture processing.* New York: Academic Press, 1976.

4

Algorithms

GEOMETRIC

Reduction

Image reduction algorithms serve to represent the image with fewer pixels. Integer reductions are easily accomplished by pixel and line deletions; for example, to reduce a digital image by a factor of three, one deletes two out of every three pixels on each line and two out of every three lines. Alternatively, integer reductions can be achieved by averaging; hence, to reduce a digital image by a factor of nine, each block of 9×9 pixels is replaced by a single pixel carrying a gray level that is the average of the 81 pixels' gray levels.

Noninteger reductions can be obtained by bilinear interpolation. After choosing the scale reduction factor, f, the gray level of any pixel (I,J) in the output image is found by:

1. Let $I' = f \times I$, truncated. Hence if $f \times I = 252.37$, then $I' = 252$. Sinilarly let $J' = f \times J$, truncated.
2. Let $Y = f \times I - I'$ and $X = f \times J - J'$
3. Let $G1$ be the gray level in the input image at point I',J'. Similarly, let

$G2$ be the gray level at the point $I',J' + 1$; $G3$ at $I' + 1, J'$ and $G4$ at $I' + 1, J' + 1$.

4. Now let $G5 = X(G2 - G1)$ and let $G6 = G3 + X(G4 - G3)$.
5. Finally, the gray level at point I,J in the output image, $G7$, is given by $G7 = G5 + Y(G6 - G5)$.

EXAMPLE:

To compute the gray-level values in the output image at $I = 21$, $J = 17$ with a scale reduction factor, f, equal to 2.4, we find $I' = 50$, $J' = 40$, $Y = .4$ and $X = .8$; by inspecting the input image we find that $G1$, the gray-level value at (I',J'), is 20. $G2$, the gray-level value at (50, 41) is 30. $G3$ is 25 and $G4$ is 40. Then $G5 = 28$, $G6 = 37$ and $G7 = 31.6$, rounded to 32. Hence the gray-level value in the output picture at $I = 21$, $J = 17$ is $G7 = 32$.

Enlargement

Digital image enlargement adds pixels to the input image to increase the scale of the output image either to improve the scale of a display for interpretation or to match the scale of another image. Just as pixel and line deletion offers the simplest form of image reduction, pixel and line replication offers the simplest form of digital image enlargement. To enlarge a digital image by an integer factor, n, each pixel is replaced by an $n \times n$ block of pixels, all with the same gray level as the replaced pixel. Figure 4.1 shows an example of replication enlargement. This form of enlargement is characterized by visibly square tiles of pixels in the output display.

Bilinear interpolation can be employed to enlarge images just as it is used

				41	41	41	47	47	47	52	52	52	53	53	53
41	47	52	53	41	41	41	47	47	47	52	52	52	53	53	53
				41	41	41	47	47	47	52	52	52	53	53	53
				44	44	44	49	49	49	56	56	56	67	67	67
44	49	56	67	44	44	44	49	49	49	56	56	56	67	67	67
				44	44	44	49	49	49	56	56	56	67	67	67
				48	48	48	55	55	55	60	60	60	70	70	70
48	55	60	70	48	48	48	55	55	55	60	60	60	70	70	70
				48	48	48	55	55	55	60	60	60	70	70	70
INPUT WINDOW							OUTPUT 3× ENLARGEMENT								

FIGURE 4.1. An example of replication enlargement.

for reduction as described above. For enlargement, however, the reduction factor, f, takes fractional values. Figure 4.2 shows part of a Landsat frame of Iran. Figure 4.3 shows the enlargement by a factor of 2 ($f = .5$) of the lower center of Figure 4.2.

Squaring

Landsat MSS imagery presents a special case requiring anisotropic enlargement. There are 3240 pixels per line of a Landsat frame, but only 2340 lines in a frame. Presenting this raw image array to a display device will result in a rectangular image with a 1.38:1 aspect ratio. This image, however, represents

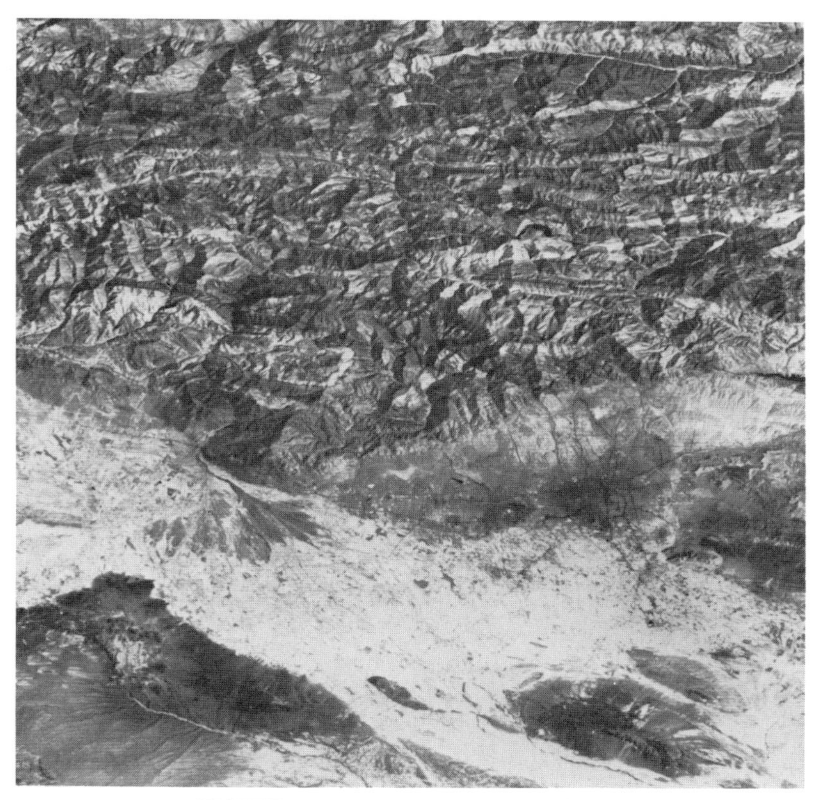

FIGURE 4.2. Part of a Landsat frame of Iran.

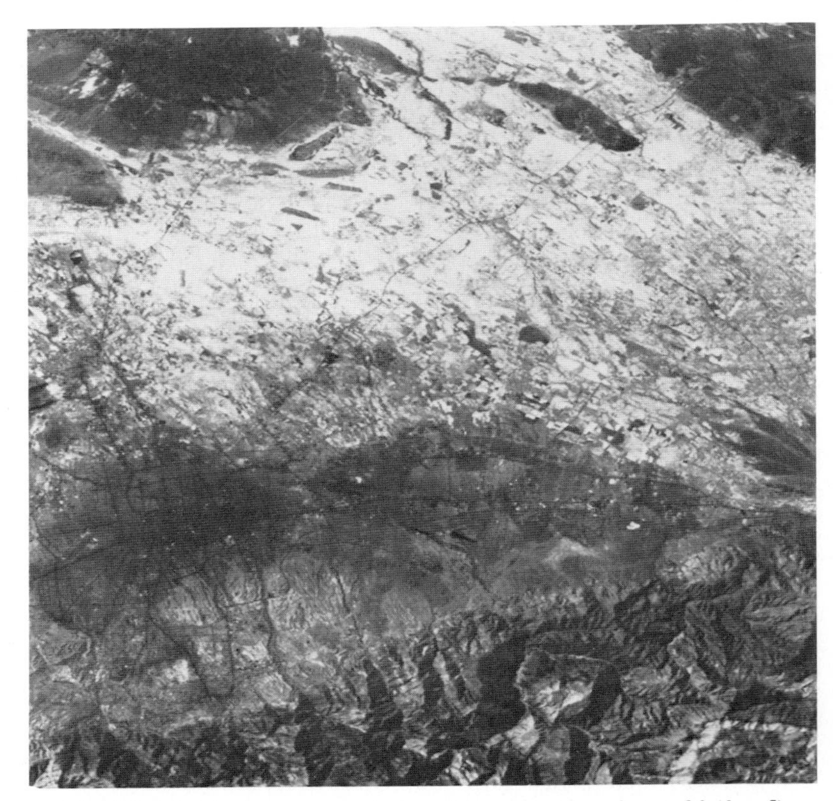

FIGURE 4.3. The lower center of Figure 4.2, enlarged by a factor of 2 (f = .5).

a square 100 nautical mile (185 km) × 100 nautical mile area on the ground. Hence a scale adjustment in either the horizontal or vertical direction is indicated.

One approximate but simple and inexpensive solution to this problem is to insert 900 additional lines in each Landsat frame. If each line of the raw Landsat image is referred to by *O* and inserted lines are termed *N*, then the scale adjusted output image will have lines sequenced as:

OOONOOONOONOOONOONOOONOOONOONOOONOONOOONOOONOONOOONOON. . . .

Said differently, after the first 3 old lines, 1 new line is inserted; after the next 3 old lines, another new line is inserted; then after 2 old lines, 1 new line; after 3 old, 1 new, and finally, after 2 old, 1 new. Then the sequence is repeated.

For each 13 old lines, 5 new lines are inserted. There are 180 such sequences in a Landsat frame. Since 5 new lines are inserted per sequence, a total of 900 new lines are added per frame. The output image has 3240 pixels per line and 3240 lines.

The gray levels of the pixels in the inserted lines can simply be copied from the line above, but more commonly they are the average of the line above and the line below. Figure 4.4 shows a portion of a Landsat scene before squaring, whereas Figure 4.5 shows the same data after squaring has been applied. Squaring is needed only for early format tapes; later format tapes can be obtained with geometric correction.

Deskewing

Not only do raw Landsat MSS digital images have need of squaring, they also have need of deskewing. Because Landsat MSS imagery is acquired with a MSS that builds up the image a few lines at a time rather than exposing an entire image plane simultaneously, a measurable amount of time transpires

FIGURE 4.4. A portion of a Landsat scene before squaring.

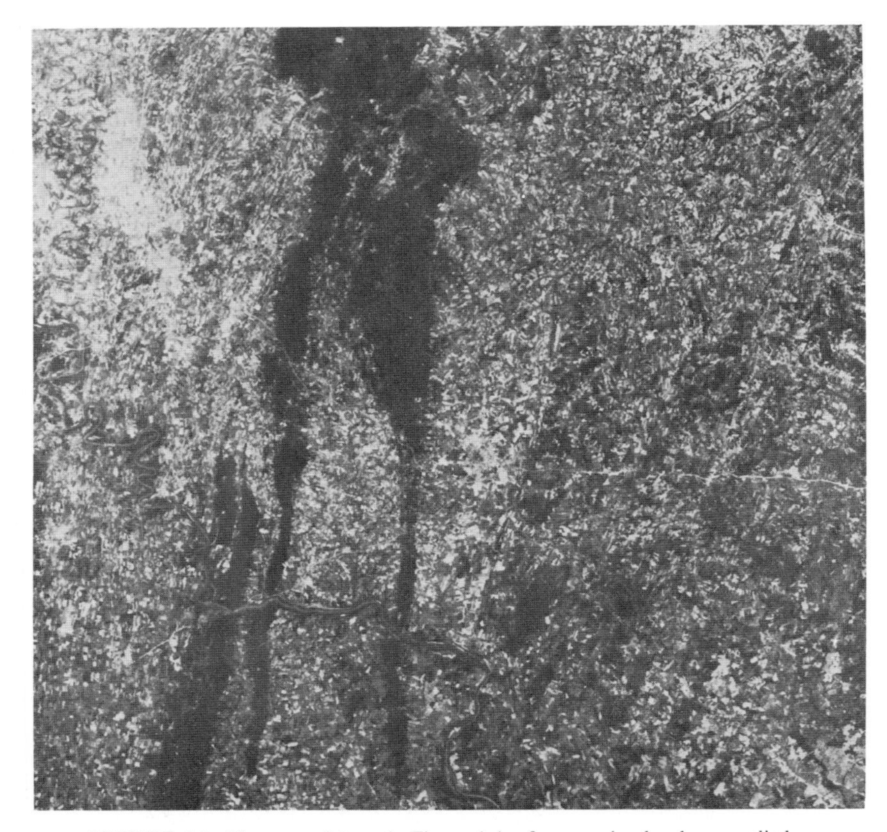

FIGURE 4.5. The same data as in Figure 4.4, after squaring has been applied.

during the acquisition of a frame. The Earth rotates under the satellite during this taking period. This introduces a skewness in the imagery. This skewness causes a north–south line on the raw imagery not to be perpendicular to an east–west line. This effect is removed by deskewing.

An approximate but simple and inexpensive deskewing can be accomplished by moving groups of lines with respect to other lines. After squaring, the Landsat frame consists of 3240 lines. These can be considered as 270 groups of 12 lines each. To deskew this image, offset each group of 12 lines 1 pixel to the left with respect to the group above it. Zeros are filled appropriately at the beginning and end of each line. The output image has 3509 pixels per line. The first group of output lines begins with 269 leading zeros whereas the last group of lines ends with 269 zeros; other groups have an intermediate number of leading and ending zeros. The geographic axes in the output image are orthogonal.

Registration

The geometric adjustments described to this point have been special cases. The general case is termed *rubber sheet stretching* and it causes each point of an input image to be relocated according to some rule. Generally this rule is expressed as a pair of transformation equations relating the new coordinates to the old coordinates through a functional expression.

When the transformation equations are chosen in a way that causes the output image to be congruent with another image or a map, the process is termed *registration* (Pratt, 1978), and image A is said to have been registered to image B. The congruency or matching is controlled by identifying landmarks that appear both on image A and image B. Commonly, these are road intersections or other features that can be located precisely on both images.

A typical set of transformation equations is termed the *affine projection*. These equations are

$$X' = (a_{11}X + a_{12}Y + a_{13})/(a_{31}X + a_{32}Y + 1)$$
$$Y' = (a_{21}X + a_{22}Y + a_{23})/(a_{31}X + a_{32}Y + 1)$$

where X,Y are the old coordinates and X',Y' are the new coordinates. There are eight coefficients to be determined in these two equations. These equations can be solved for the eight unknowns if the coordinates of four landmarks, or *control points,* as they are often called, can be determined from both image A and image B since each control point supplies data for two equations. If more than four control points are employed, the solution is overdetermined and a least squares fit is used to find the best values for the coefficients.

Since both image A and image B are planar, the affine transformation is generally sufficient. Rotation, translation, and anisotropic scale change are accommodated by this transformation. More elaborate transformations are required if a straight line in image A is to be mapped into a curved line in image B, or if higher-order corrections are sought.

In practice, the transformation equations are transposed to express X,Y as functions of X',Y', then pixel by pixel, line by line, the new image is built up. Each X',Y' is plugged into the equations to find the corresponding X,Y, *and the gray level associated with* X,Y is assigned to the new image at X',Y'. A large amount of random access memory is useful since all of the input image can then be stored in memory. The look-up of the gray level at X,Y is more efficient than searching for it on tape. The reader seeking further information on reduction and enlargement should consult Ramapriyan (1977).

Assembling Mosaics

When two or more overlapping images are to be combined into a mosaic, that is, brought into registration and then treated as a single image, the seams between the picture sections are generally apparent. This spurious artificial edge often is distracting, and can interfere with photointerpretation. These seams result from changes that occur between the times of acquisition of the image sections. For example, one may choose to make a mosaic from sections of two Landsat images. East–west adjacent Landsat scenes have about 15% overlap. But since the frames are not acquired simultaneously, substantial changes may have taken place between these acquisitions. The atmospheric transmission or cloud cover can differ in the two scenes. The soil moisture content may have changed. The vegetation may have seasonally changed. Perhaps only one of the scenes has snow cover. Even the sensor radiometric calibration may have been adjusted.

A method of generating "seamless" mosaics of digital image sections is described in Milgram (1975); an improvement on this method is described in Milgram (1977). The method has four steps. In the first step, the two image sections are brought to registration using the methods described earlier. An example of the output from this step is shown in Figure 4.6. Here the image sections were acquired 18 days apart and are rather different in appearance.

Once the overlap region has been identified on the registered images, the gray level of one of the images, L or R (left or right), is adjusted linearly (see page 76) until the average gray level in the overlap region of image L matches the average gray level in the overlap region of image R. This completes the second step: the output image still shows the seam but it is substantially less visible.

In the third step a feathered seam is created. On a line by line basis, a juncture point is chosen; the picture information in the final mosaic to the left of the juncture point will come from the line segment of image L, and image R will supply the rest of that line in the mosaic. Define

$$D(n) = \sum_{i=-(w/2)+1}^{i=w/2} |L(n+i) - R(n+i)|$$

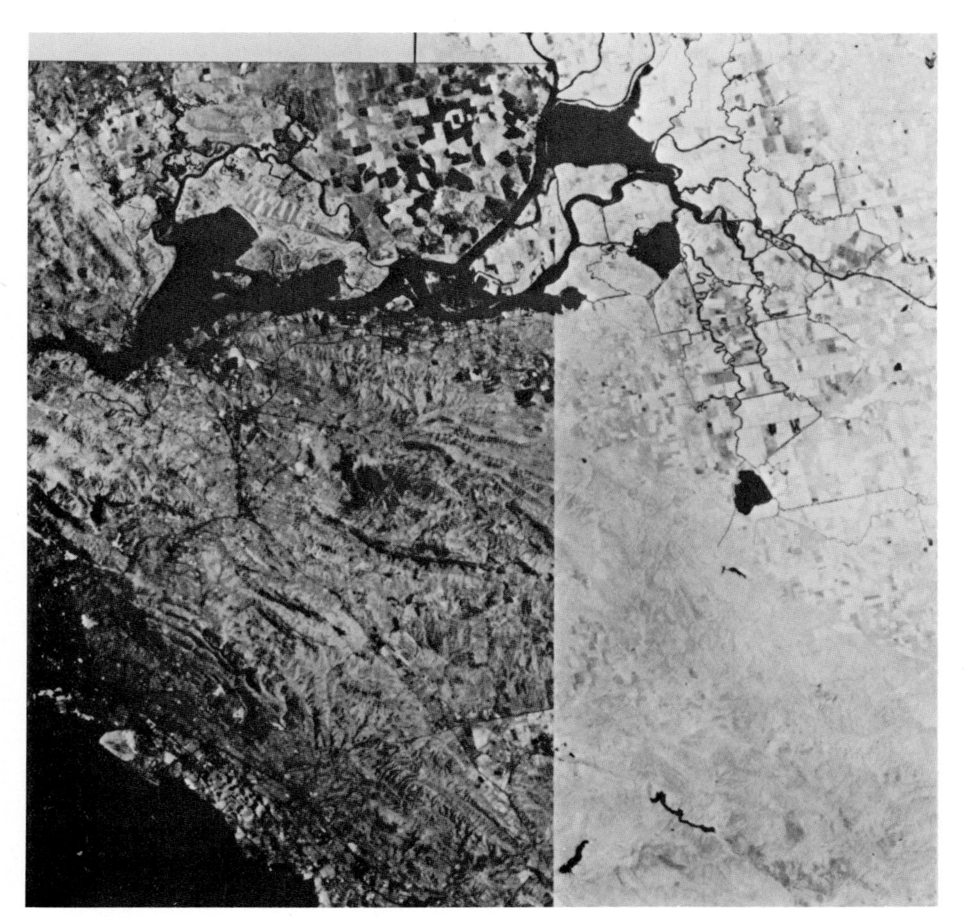

FIGURE 4.6. Sample output from first step of method of generating seamless mosaics of digital image sections. The image sections were acquired 18 days apart and are rather different in appearance.

where $L(j)$, $J = 1,2, \ldots, K$ are the gray-level values of the current line in the overlap region of image L. Similarly, $R(j)$, $j = 1, 2, \ldots, K$ for R. Choose a value, say 8, for the parameter w. Naturally, w must be less than the width of the overlap region. To find this line's juncture point, one finds the value of n that minimizes $D(n)$. This method finds a succession of juncture points, one per line, with horizontal positions that are unrelated to each other. A refinement of this algorithm constrains the juncture point of a given line to be near those of adjacent lines.

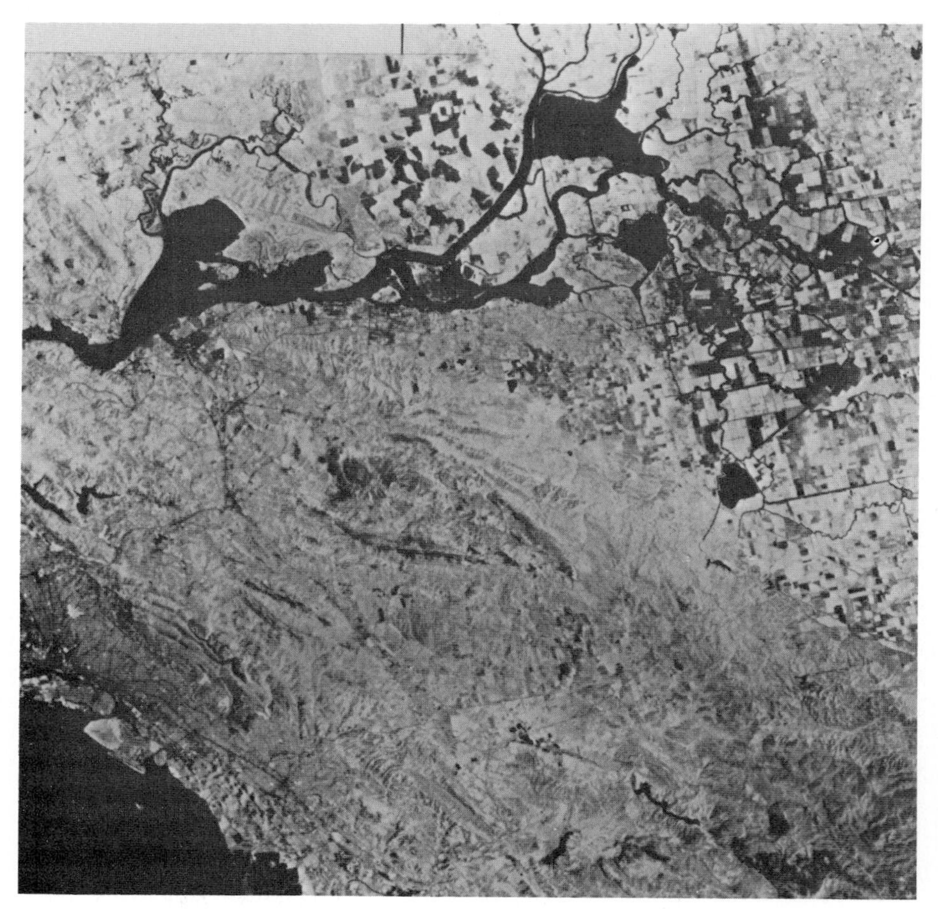

FIGURE 4.7. The final output of the method of generating seamless mosaics, applied to the same image segments shown in Figure 4.6.

The final step is the smoothing of the gray-level change at each of the juncture points. Figure 4.7 shows the output of this four-step algorithm applied to the same image segments shown in Figure 4.6.

ENHANCEMENT

Digital image enhancement is the enterprise of making good pictures from bad. This section addresses some of the methods for accomplishing this.

Unfortunately, universal agreement about what constitutes a good, or at least a better, picture has been elusive. A number of quantitative measures of image quality have been developed, but acceptance of these is not unanimous. Here, *enhanced* means more interpretable, although the techniques to be described are said by some interpreters to vary in utility. To a degree, then, enhancement must be regarded as an art (Boulter, 1979).

Tabular Gray-Scale Adjustment

One of the most frequent enhancement operations to be applied to digital imagery is gray-scale adjustment. For example, although Landsat MSS data on CCTs allow 8 bits for encoding the gray levels of the pixels, the gray levels for bands 4, 5, and 6 range only between 0 and 127. The gray levels for band 7 range from 0 to 63. To display this data effectively on an image display device with a 0–255 dynamic range requires that a gray-level adjustment be performed before data display.

The gray-scale adjustment process accepts a digital image as input and produces an output image in which each of the pixel gray levels has been changed according to some rule. One basic scheme is a table look-up process that evaluates the output gray level for a given pixel on the basis of the input gray level for that pixel and a correspondence specified in a table.

These tables are constructed from a priori knowledge, as in the Landsat case discussed above, or are developed experimentally, based on experience of what adjustment has been effective for the application at hand. Inspection of the histogram can facilitate the choice of table values.

Slicing and Gray-Scale Reversal

Gray-scale adjustment rules can also be specified algebraically. One gray-scale adjustment, in such common use that it has been named *slicing,* can be specified as follows:

$$G' = 0;\ G = 0,1,2,\ \ldots\ ,K$$
$$G' = G;\ G = K + 1,\ K + 2,\ \ldots\ ,M$$
$$G' = 0;\ G = M + 1,\ M + 2.\ \ldots$$

The user sets the thresholds K and M. G is the input gray level and G' is the output gray level.

Gray-level reversal can also be expressed in a program algebraically, as follows:

$$G' = 255 - G$$

where it has been assumed that both the input and output gray levels range from 0 to 255 (8-bit coding). Figure 4.8 shows a section of a Landsat scene over the New York Bight, and Figure 4.9 shows the same scene with a reversed gray scale and sliced.

Linear Stretch

Another special form of gray-scale adjustment, called *stretching,* is expressed in a program algebraically, as follows:

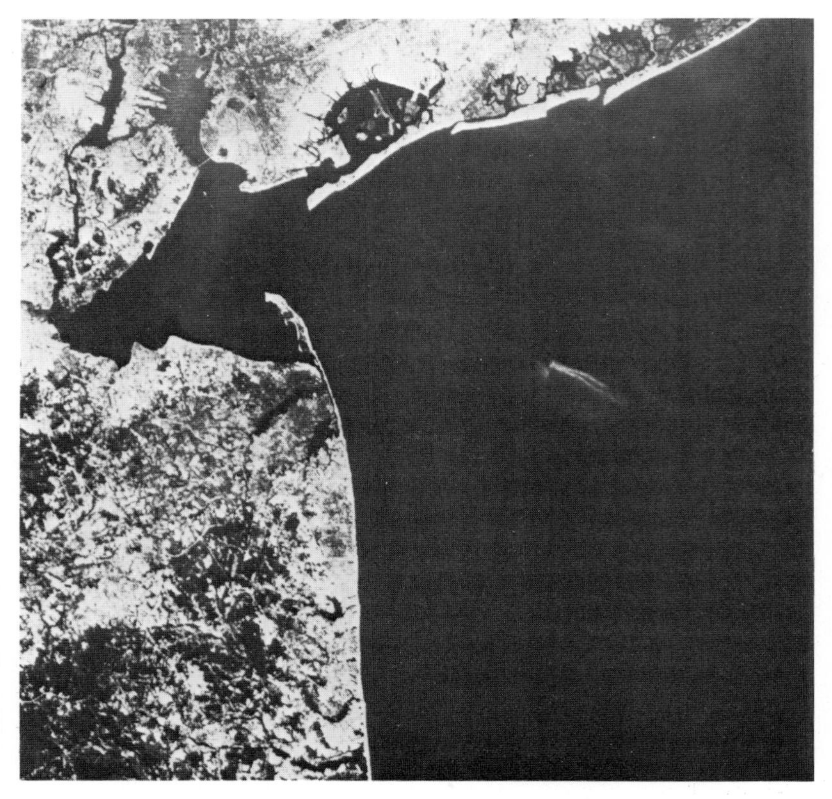

FIGURE 4.8. A section of a Landsat scene over the New York Bight.

$$
\begin{aligned}
G' &= 0;\ \mathrm{G} = 0, 1, 2, \ldots, K \\
G' &= A \times G + B;\ G = K + 1, K + 2, \ldots, M \\
G' &= 255;\ G = M + 1, M + 2, \ldots, 255.
\end{aligned}
$$

The parameters A and B are set:

$$
\begin{aligned}
A &= 255/(M - K) \\
B &= -K \times A.
\end{aligned}
$$

Again, this assumes an output gray-level range of 0–255 (8-bit coding). This linear stretching allows a digital image of restricted gray-level range to use the full dynamic range of the display device.

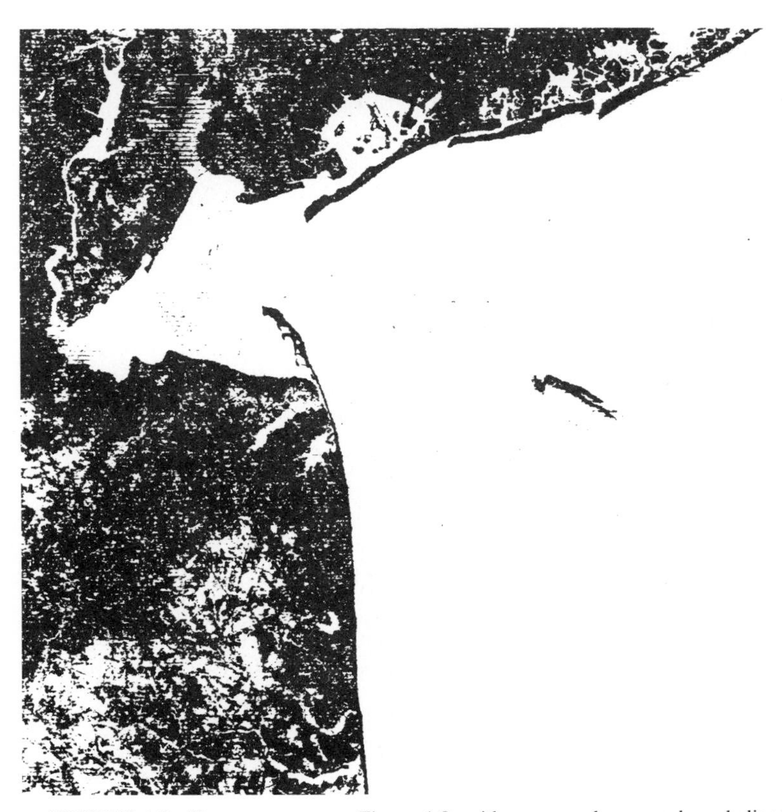

FIGURE 4.9. The same scene as Figure 4.8, with a reversed gray scale and sliced.

Histogram Equalization

Perhaps the most satisfying gray-scale adjustment algorithm is termed *histogram equalization* (Pratt, 1978; Rosenfeld & Kak, 1976). This is a nonlinear gray-scale adjustment that has demonstrated its effectiveness in making digital images more interpretable.

For an image having N gray levels, let the normalized histogram be $h(n)$, $n = 0, 1, \ldots, N - 1$. Define the cumulative histogram to be

$$H(n) = \sum_{i=0}^{n} h(i).$$

$H(n)$ is an increasing function that ranges from 0 to 1. The adjustment rule then is

$$G' = H(n) \times G, \text{ for } G = n.$$

This assumes that the input image and the output image have the same dynamic range.

Generally, this is implemented as a special case of the tabular gray-scale adjustment algorithm discussed earlier. A subroutine computes the cumulative histogram and loads it into the gray-scale adjustment table storage space and, after multiplying each table entry by n, proceeds to apply the tabular gray-scale adjustment algorithm described on page 75.

The histogram of the resulting adjusted output image is much flatter; the count of dark pixels is found to be equal to the count of light pixels. Additional information on histogram enhancement can be found in Amin (1977), Frei (1977), and Hummel (1977).

Edge Enhancement

As described earlier, an image can be considered a mathematical function of two variables. Symbolically, $Z = f(X,Y)$. Accordingly, conventional mathematical operations can be performed on this picture function. One particularly useful operation is an isotropic second derivative, the Laplacian. Symbolically, the Laplacian is given by:

$$L = (d^2f/dX^2) + (d^2f/dY^2)$$

To find the Laplacian of a digital image, finite differences are used to estimate the derivatives indicated. Let $G(I,J)$ represent the gray level at the point I,J in a digital image. Then the digital image Laplacian is:

$$L(I,J) = G(I - 1,J) + G(I + 1,J) + G(I,J - 1) + G(I,J + 1) - 4G(I,J)$$

This quantity takes on high values wherever the rate of gray-level change itself changes. High values occur at edges in the original picture. An example of a Laplacian as applied to a Landsat view of Honey Lake, California is shown in Figure 4.10.

The edges of a digital image are considerably sharpened by adding the Laplacian to the original image (Pratt, 1978):

$$G'(I,J) = G(I,J) + K \times L(I,J)$$

where K is a weighting factor. Figure 4.11 is a portion of a Landsat scene in Iran before edge enhancement. The Laplacian-added, edge-enhanced version of this scene is shown in Figure 4.12.

Smoothing and Convolution

Some digital images have an unacceptable amount of noise. This shows up as stripes, herringbone patterns, salt and pepper, haze, or any number of other

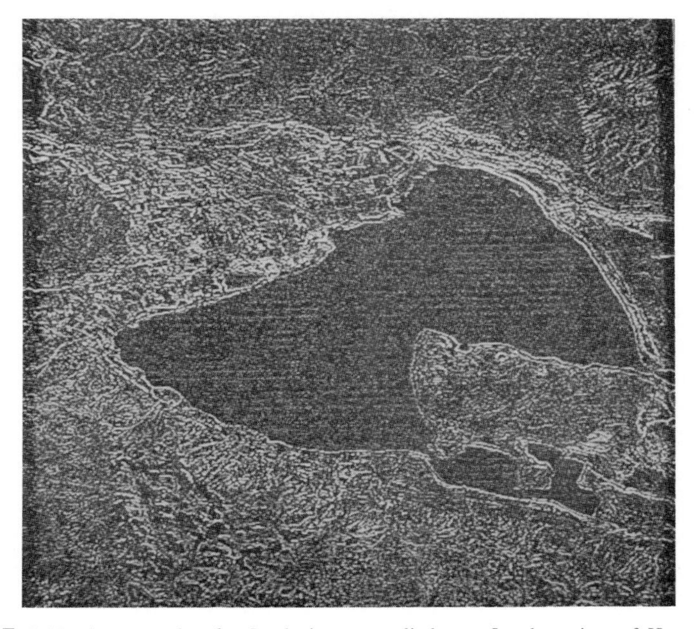

FIGURE 4.10. An example of a Laplacian as applied to a Landsat view of Honey Lake, California.

FIGURE 4.11. A portion of a Landsat scene in Iran before edge enhancement.

effects. This can obscure image content and diminish the utility of an image. The most effective way to compensate for noise is to characterize the statistical nature of the noise, perhaps even identify the source of the noise and deterministically quantify the noise component, and design the remedial operations for the specific case at hand. Often this is impractical.

Smoothing can be a plausible compromise. *Smoothing* is a generic term that includes a wide variety of operations. Their common purpose is to restrain extreme fluctuations in gray level from point to point in the image (Panda, 1978). One scheme is to scan through the image, pixel by pixel, adjusting gray-level values to assure that the change in gray-level value from one pixel to the next never exceeds some maximum allowable amount.

Another scheme is to apply sequentially two of the algorithms described

FIGURE 4.12. The Laplacian-added, edge-enhanced version of the scene in Figure 4.11.

earlier. First execute an image reduction, say by a factor of 2, using the averaging algorithm discussed at the beginning of this chapter. Then enlarge the resulting image by the same factor, using the bilinear interpolation algorithm.

Smoothing can also be achieved as a special case of a more general operation termed *convolution* (Pratt, 1978; Rosenfeld & Kak, 1976). Two-dimensional convolution requires the specification of an operator, also called a *kernel*. This kernel consists of an array of coefficients. This array commonly measures 3 × 3, 5 × 5, or 7 × 7. To obtain the value of the convolution of an image A with a kernel K at image point I,J, one imagines the kernel superimposed on the image with the center point of the kernel located at the image point I,J. Each of the kernel coefficients is multiplied by the gray level of the image point with which it is associated. These multiplicative products are then added together. This is evaluated at each point of image A and the output image B is changed from A in ways that depend on the coefficients in the kernel. Symbolically, $B = K * A$.

The identity kernel, so that $B = A$, is a one at the center with all other coefficients zero. The Laplacian kernel that results in the Laplacian discussed in edge enhancement is a 3 × 3 kernel:

$$\begin{matrix} 0 & 1 & 0 \\ 1 & -4 & 1 \\ 0 & 1 & 0 \end{matrix}$$

A kernel with all coefficients equal to $1/Q$, where Q is the number of coefficients in the kernel, produces effective smoothing.

Fourier Filtering[1]

In the realm of general digital image processing, two-dimensional Fourier transforms are used for enhancement, compression, texture classification, smear removal, quality assessment, cross-correlation, and a host of other operations. One simple example is contrast improvement. Here the Fourier transform of an input image is obtained to ascertain the spatial frequency content of that image. If the zero-frequency component of that transform is set to zero, the inverse of the "filtered" transform will be a new rendering of the image with the background haze removed. Hence the contrast is improved and the image will be more interpretable.

Other spatial filtering to modify an image or to analyze a spatial structure can often be performed conveniently in the frequency domain. Recent improvements in the bandwidth of conventional hardware fast Fourier transform (FFT) processors could increase the importance of frequency domain image processing and analysis.

There is an extensive literature describing digital frequency domain image processing: texts by Andrews (1970), Lipkin and Rosenfeld (1970), and Rosenfeld (1969); surveys in Huang, Schreiker, and Tretiak (1971), Andrews (1974*b*), Hale and Saraga (1974), Nagy (1972), and Rosenfeld (1972); articles by Billingsley (1970), Herbst and Will (1972), Hunt (1972), Jarvis (1974), and O'Handley and Green (1972). Two-dimensional digital filters are described in Frieden (1972), Hall (1972), Helstrom (1967), Hunt (1973),

[1]This section adapted from Boulter (1976).

Lahart (1974), McGlamery (1967), Richardson (1972), and Sondhi (1972). FFT filters and general techniques can be found in Andrews (1974*a*), Bergland (1969*a*, 1969*b*, 1969*c*), Buijs, Pomerleau, and Fournier (1974), Champeney (1973), Cochran *et al.* (1967), Cooley, Lewis, and Welch (1967, 1969), Corinthios, Smith, and Yen (1975), Gallagher and Liu (1974), Gottlieb and Lorenzo (1974), and Helms (1967).

The Fourier transform of a function $f(x,y)$ that has been sampled over an $N \times N$ grid of points is given by:

$$F(u,v) = \sum_{x=0}^{N-1} \sum_{y=0}^{N-1} f(x,y) \exp[-2\pi i(ux + vy)/N] \tag{1}$$

and the inverse by:

$$f(x,y) = \sum_{u=0}^{N-1} \sum_{v=0}^{N-1} F(u,v) \exp[2\pi i(ux + vy)/N] \tag{2}$$

Rearrangement of equation into the form:

$$F(u,v) = \sum_{y=0}^{N-1} \sum_{x=0}^{N-1} f(x,y) \exp(-2\pi iux/N) \exp(-2\pi ivy/N) \tag{3}$$

shows that the two-dimensional transformation can be performed by applying one-dimensional transforms to each of the rows (or columns) of the input array and then similarly transforming the columns (or rows) of the intermediate array.

Expanding the complex exponentials into sine and cosine terms and assuming that $f(x,y)$ is real allows Eq. (1) to be decomposed into four components:

$$F(u,v) = (A - B) - i(C + D) \tag{4}$$

where

$$A = \Sigma f(x,y) \cos(u'x) \cos(v'y)$$
$$B = \Sigma f(x,y) \sin(u'x) \sin(v'y)$$
$$C = \Sigma f(x,y) \sin(u'x) \cos(v'y)$$
$$D = \Sigma f(x,y) \cos(u'x) \sin(v'y).$$

The sigma refers to the double summation performed in Eq. (1) and $u' = 2\pi u/N$, $v' = 2\pi v/N$.

Successive application of a standard one-dimensional FFT algorithm (compacted for use with real functions and with an output in the form of interleaved sine and cosine components) to the rows and columns of the input array yields an $N + 2 \times N + 2$ element array of coefficients in which the four components of Eq. (4) are grouped. The zero-frequency term is located in one corner of the array and the orthogonal spatial frequencies u and v span the two axes up to the limiting Nyquist frequencies U and V. Since row 2 and column 2 (which contain the term sin 0), and row $N - 1$ and column $N - 1$ (which contain the term $\sin\pi$) are identically zero, the two-dimensional transform can be compressed to fit into the original $N \times N$ input array.

This format for storing the two-dimensional Fourier transform of a real image, in which the four terms corresponding to each spatial frequency are grouped together rather than being located in four separate quadrants, has the advantages of simplifying certain frequency domain filtering operations and of providing a gray-scale display that, if the absolute values of the coefficients are used, resembles the power spectrum.

A display of the power spectrum (anologous to a conventional optical diffraction pattern) in which the zero-frequency component is located at the center of the image and the two axes span the range between plus and minus the Nyquist frequencies, can be obtained from the preceding array. Using an asterisk to represent the complex conjugate, the power spectrum is defined as:

$$S(u,v) = F(u,v)\, F^*(u,v)$$

This power spectrum can be used for image restoration.

The goal of restoration is to process a degraded image so that, in accordance with some criterion, it resembles as closely as possible some ideal image. Only limited progress has been made in relating an objective restoration criterion to the optimum criterion for a human observer.

Restoration such that the mean-square error between the restored and the ideal image is minimized represents one possible objective approach. This approach is adopted here in two forms to deconvolve a shift-invariant point-spread function in the presence of additive white noise.

Assume that an image with power spectrum $S_f(u,v)$ has been degraded by a point-spread function with Fourier transform $H(u,v)$ and that uncorrelated noise with power spectrum $S_n(u,v)$ has been added to the blurred image. Helstrom (1967) has shown that the linear shift-invariant filter that yields a restored image with the least mean deviation from the ideal image is given by:

$$Q(u,v) = \frac{H^*(u,v)\, S_f(u,v)}{|H(u,v)|^2 S_f(u,v) + S_n(u,v)}$$

When the signal:noise ratio is small, the filter smooths to reduce noise without performing much high-frequency enhancement, whereas for higher signal:noise ratios, more sharpening occurs.

This restoring filter requires estimates for the power spectra of the ideal image and of the noise. One approach is to set the ratio $S_f(u,v):S_n(u,v)$ to a constant equal to the average signal:noise power in the image. An alternate method is advantageous if there are regions of high signal:noise ratio in the power spectrum of the degraded image. This can be due, for example, to the transformation of a particular spatial structure to small areas in the frequency domain. In this case, use of the power spectrum of the degraded image as an estimate for $S_f(u,v)$ yields preferential restoration of that particular spatial structure.

Destriping

Landsat MSS images frequently suffer from striping, an image degradation characterized by an overall pattern of horizontal stripes. The cause of this problem is the radiometric calibration of the Landsat MSS sensors. The MSS image is acquired six lines at a time; there are six sensors per MSS spectral band. When the radiometric sensitivity of any of the sensors is unbalanced with respect to the other sensors for that band, the image lines generated by that sensor are noticeably brighter or dimmer than the other lines of the image and a striped pattern results.

To correct this problem, the first step is to obtain a set of six histograms for the image, one for each of the sensors that acquired the image. The first histogram is generated from the pixel gray-level values from lines 1, 7, 13, 19 . . . ; the second histogram pertains to lines 2, 8, 14, 20 . . . ; the third describes the pixels in lines 3, 9, 15, 21 . . . ; and so forth. Comparison of these histograms will in most cases quickly indicate the misbehaving set of image lines.

This striping is then removed by applying a tabular gray-scale adjustment only to the pixels on the stripe lines: the pixels on the other lines are not altered. The gray-scale adjustment table to be used in the adjustment process is chosen to produce a histogram for these stripe lines that resembles the histograms associated with the other sensors (Horn & Woodham, 1978).

This section has dealt briefly with some of the problems with and techniques for approaching computer enhancement. For additional reading on this subject, the interested reader is referred to Andrews and Hunt (1977), Kreshaven and Srinath (1977), Lev, Zucker, and Rosenfeld (1977), and Panda and Kak (1977).

MULTIPLE IMAGE OPERATIONS

Addition and Subtraction

Image addition and subtraction are quite common operations. The discussion of edge enhancement earlier touched on image addition; the edge-enhanced image is obtained by adding the ''image'' termed the *Laplacian* to the original image. To perform image addition, the gray levels of the output image C are set equal to the sum of the corresponding gray levels in image A and image B. Symbolically, $G_C = G_A + G_B$.

Care must be taken in image addition that image A and image B are in registration, that they have the same format size, and that the largest value of the individual pixel sums do not exceed the highest allowed gray level. Image subtraction is obviously quite similar. The major change is that for image subtraction one must take care that the smallest result from the individual pixel subtractions must remain nonnegative.

More generally, pairs of images may be linearly combined as follows:

$$G_C = m \times G_A + n \times G_B + C.$$

Where m, n, and C are constants to be specified by the user.

Ratios

A process that sometimes produces a usefully different presentation of multiband images data is the ratio image. This consists of a computed image with the gray level of each pixel equal to the gray-level ratio of the corresponding pixels in two bands of a multiband input image. Conceptually,

$$G_C = G_A/G_B.$$

Unfortunately, this simple form of image ratio is not very effective for several reasons. Since $G_B = 0$ is possible, G_C is unbounded. Additionally, whenever G_B is greater than G_A, then G_C is less than 1, and thus perhaps one-half the pixels are mapped into gray-level zero.

A better algorithm is

$$G_C = R \arctan(G_A/G_B)$$

Here G_C ranges from zero to $1.57R$; for $R = 162.34$, G_C ranges from zero to 255. This also provides a better distribution of G_C values since the occurrences of G_B larger than G_A are mapped into 128 possible gray-level values for G_C as desired. Although a pseudo-ratio, this and other variants including logarithms are commonly termed simply *ratio*.

One reason this process is useful is that G_C is not dependent on variations in illumination in different parts of the scene. For example, some sensors have a characteristic, termed *vignette,* which causes pixels at the edge of the format to receive less light than pixels at the center. But since vignette is not wavelength dependent, G_C for a given land cover type will have the same value at the center as it has at the format edge.

Figure 4.13 shows part of band 5 of an Oklahoma Landsat scene; the same area as imaged in band 7 is shown in Figure 4.14. Figure 4.15 is the ratio of the band 5 image to the band 7 image. Another ratio image that assists in crop field delineation is shown in Chapter 5, page 157.

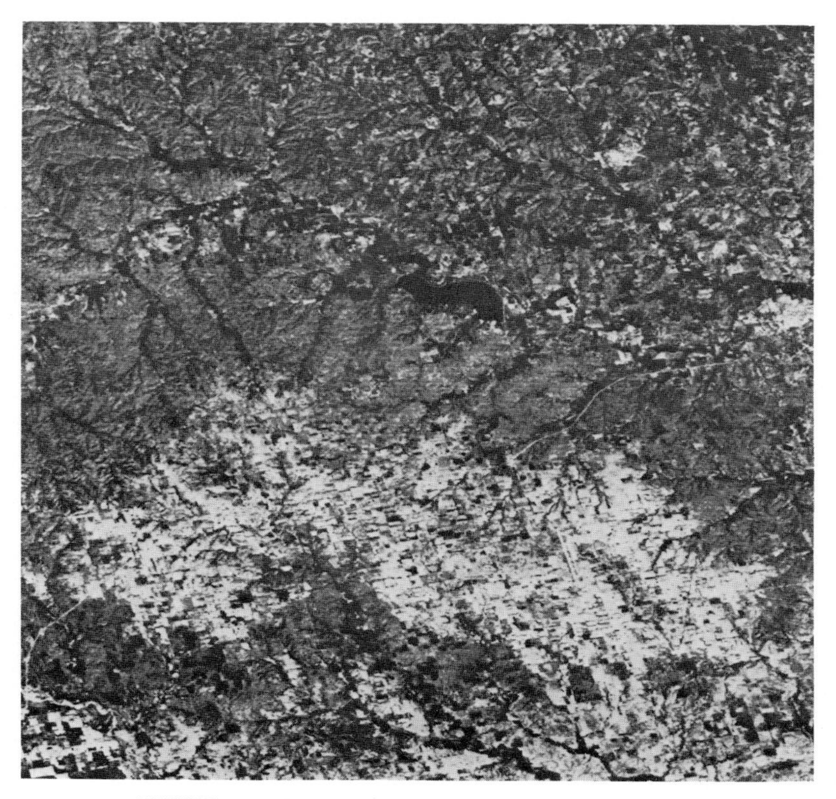

FIGURE 4.13. Part of band 5 of an Oklahoma Landsat scene.

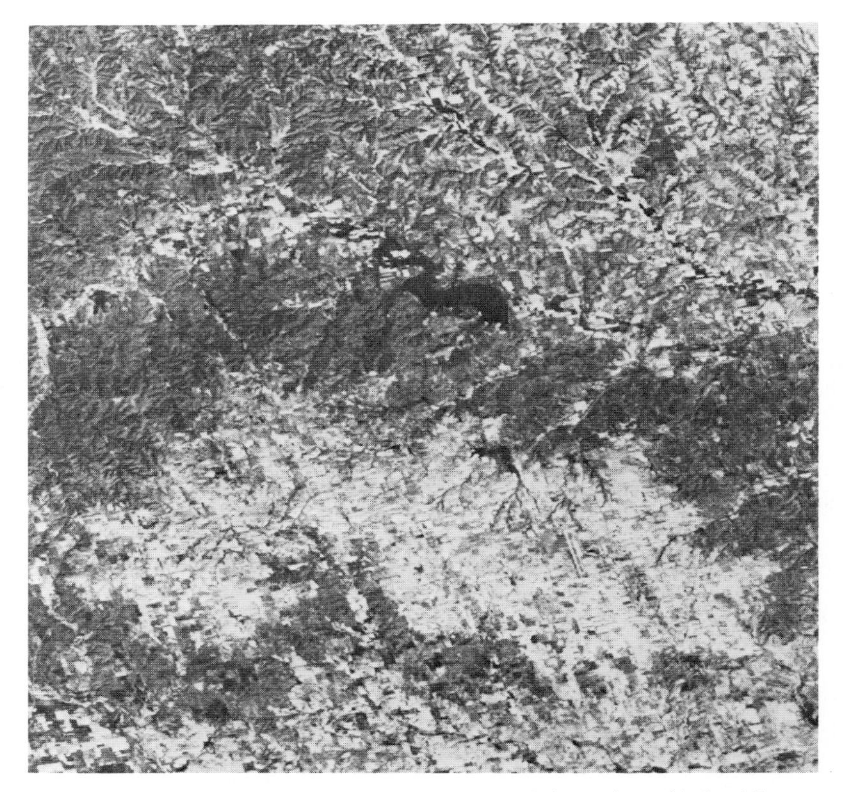

FIGURE 4.14. The same area shown in Figure 4.13, but as imaged in band 7.

Template Matching

An important activity in satellite navigation and other applications is the recognition and location of landmarks on images of the earth's surface, as viewed by the satellite. Given the locations of several landmarks on the image and their true geographical locations, the position and orientation of the satellite can be determined.

We shall assume here that the approximate position and orientation of the satellite is already known and that we want to use the landmark data to obtain more accurate estimates. This implies that we know approximately where, on the image, landmarks should appear—perhaps to within a few dozen pixels. We shall ignore here any errors in our estimation of the orientations and sizes

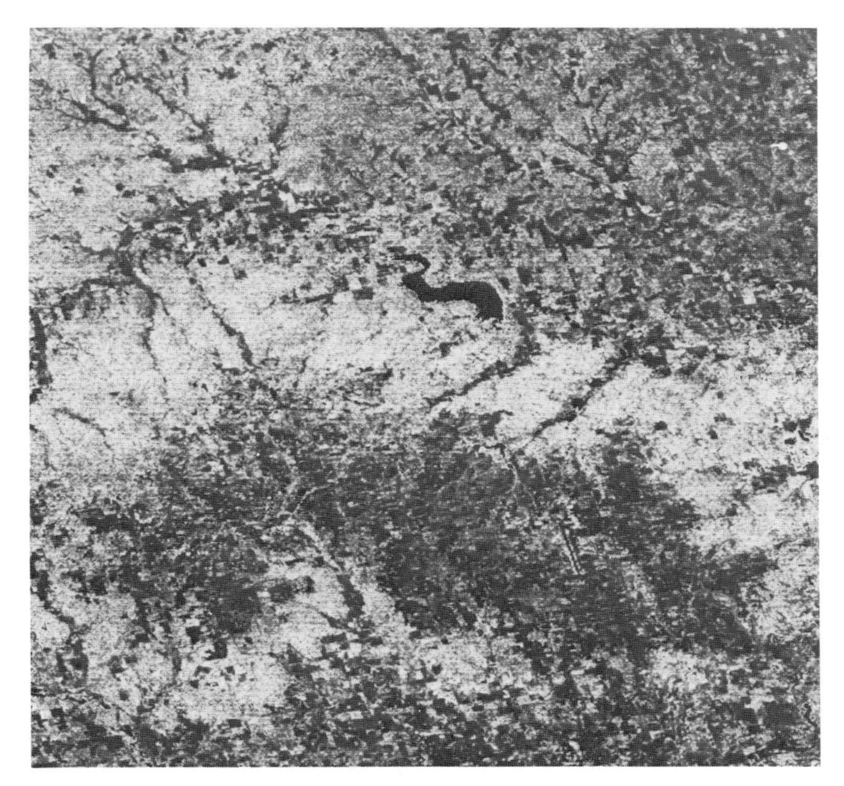

FIGURE 4.15. The ratio of the band 5 image to the band 7 image.

of the features; it is reasonable to assume that if the error in estimated position is small, then the errors in estimated orientation and size are negligible.

If we regard the orientations and sizes of the landmarks on the image as known and their positions as approximately known, the problem of landmark identification and location can be solved by a template matching process. This basically involves comparing a template, or small image of the landmark, with the full picture in the range of positions where the landmark could be located (Shapiro & Haralick, 1980). If a good match is obtained, the landmark has been detected, and the position of this good match gives the location of the landmark.

There are many different template matching processes that could be used for this purpose, and the matching can be implemented in various ways (digitally, optically, or by a human operator) (Pratt, 1978; Rosenfeld & Kak,

1976). We shall consider here only automated digital techniques. One goal of this effort is to make the template matching as computationally efficient as possible, while keeping machine storage requirements at the minicomputer level.

The standard digital template matching technique, which is based on cross-correlation, is relatively slow. Substantially faster results can be obtained by using an absolute difference measure of mismatch (sometimes known as the SSDA, or Sequential Similarity Detection Algorithm). This approach is not only computationally less expensive but also lends itself to the use of efficient search techniques for speeding up the detection of mismatches, as shown by Nagel and Rosenfeld (1972). It can be combined with various pre- and postprocessing techniques to increase the sharpness of the matches obtained, thereby increasing the accuracy, as well as the speed, with which landmarks can be located.

The most commonly used measure of the match between a template, $t(x,y)$, and a picture, $p(x,y)$, is the correlation coefficient (or normalized cross-correlation)

$$\frac{\iint_T t(x,y)\, p(x+u, y+v)\, dxdy}{\sqrt{\iint_T t^2(x,y)\, dxdy \iint_T p^2(x+u, y+v)\, dxdy}}$$

where (u,v) is the displacement of the template relative to the picture. Here the area of integration, T, is the area occupied by the template. This coefficient takes on values between 0 and 1 and has value 1 only when $t(x,y)$ and $px + u, y + v)$ are identical (over the area T) except for a multiplicative constant. In the digital case, the expression for the correlation coefficient is identical, except that the integrals are replaced by sums over picture element arrays.

Computation of the correlation coefficient is relatively slow since it involves a large number of multiplications of template values by picture values for each position of the template relative to the picture. The process can be performed faster by using Fourier transform techniques to compute the cross-correlation; however, this approach is costly in computer memory requirements and will not be discussed further here.

A more quickly computable measure of the mismatch between a template and a picture is the integrated (or summed) absolute difference:

$$\iint_T |t(x,y) - p(x+u, y+v)|\, dxdy.$$

This measure is zero when t and p are identical (over the area T). Computation of this measure involves no multiplications; many fewer arithmetic operations are required than in the case of the correlation measure. In addition, use of the absolute difference measure makes it possible to take advantage of shortcuts in the matching process.

A major advantage of the sum of the absolute differences as a mismatch measure is that this measure grows monotonically with the number of points whose absolute differences have been added into the sum. This means that if we are looking for a point of best match in a given region, we can stop adding up absolute differences in a given position (u,v) as soon as the sum has grown larger than the smallest sum so far obtained in any previously evaluated position (u_0,v_0), since we now know that (u,v) cannot possibly be the position of best match.

We can do even better if we set up an absolute threshold such that if the mismatch measure exceeds θ at a given point, we will not accept that point as a point of match. It is now sufficient to add up terms of the sum until θ is exceeded; when this happens, we can reject the given point and move on to the next point. By using both mismatch comparison and absolute thresholding, the process of finding a best match (below the mismatch threshold) can be made considerably faster. For points where the match is poor, the threshold should be exceeded rapidly so that only a fraction of the terms in the sum need be computed for such points. Even for points where the match is good, we need not always compute all the terms of the sum since we can stop as soon as we exceed the smallest previously obtained sum.

It has been pointed out by Nagel and Rosenfeld (1972) that the expected amount of time required to exceed a mismatch threshold can be reduced by matching template pixels against the corresponding image pixels in a preselected order. To illustrate this idea, let us consider a simple example. Suppose that the template and image contain only three gray levels, say 0, 1, and 2, and suppose that, in the given region of the image, these levels occur with relative frequencies $\frac{1}{2}$, $\frac{1}{3}$, and $\frac{1}{6}$, respectively. If the template is not at or near the correct match position, we can assume that the image gray level that coincides with a given point of the template is uncorrelated with the template gray level at that point. Thus, we can compute the expected absolute gray-level difference between template and image at a point as shown in Table 4.1.

In other words, the expected absolute difference between template and image at a template point having gray level 0 or 1 is $\frac{2}{3}$; but at a template point of gray level 2, the expected absolute difference is $\frac{4}{3}$. Thus, we can expect,

TABLE 4.1
Computation of Absolute Gray-Level Differences

Template gray level	Probability of difference					Probability of absolute difference			Expected absolute difference
	−2	−1	0	1	2	0	1	2	
0	1/6	1/3	1/2	0	0	1/2	1/3	1/6	2/3
1	0	1/6	1/3	1/2	0	1/3	2/3	0	2/3
2	0	0	1/6	1/3	1/2	1/6	1/3	1/2	4/3

on the average, a faster rate of growth of the sum of absolute differences if we compute the difference first for template points that have gray level 2. Thus the mismatch threshold (or the lowest level of mismatch previously found) is likely to be exceeded sooner if we compare template points with image points in a special order—namely, first using template points of gray level 2—rather than using the template points in an arbitrary order.

Generalizing the example just given, it is easily seen that the mismatch threshold (absolute or relative) will be exceeded faster, on the average, if the template points are compared with the corresponding image points *in decreasing order of expected absolute gray-level difference*. The degree of speed-up achievable in this way depends, of course, on the probability distribution of gray levels in the given region of the image; if all gray levels are equally likely, no speedup is possible by this method.

The method just described is even more effective when we are matching multispectral, rather than monochromatic, gray-scale images. This is because a probability distribution of multispectral vector values is even more nonuniform than a distribution of gray levels. Thus, it is possible to select template points whose expected absolute differences from the corresponding image points should be quite large. Here the ''differences'' must, of course, be defined in vector terms, for example, as sums of absolute differences of the individual vector components.

The speed-up obtainable by this method will not be the same for all landmarks. But the best results should be obtained for landmarks in whose vicinity the distribution of gray levels, or multispectral values, is highly nonuniform. Given a set of available landmarks, one can determine, for each of them, how much speed-up can be expected and one can select preferred landmarks for which the expected speed-up is greatest.

The position of ''best'' match, determined by any method, may not be sharply defined, since there may be a set of hearby positions for which the match is nearly as good. It has been known for years that match sharpness can be improved by differentiating, or high-pass filtering, the image and the template before matching them. In other words, outlines of regions give sharper matches than do solid regions.[2] It would be of interest to investigate whether the matching process described previously is improved by using differentiated or outlined images and templates. This will certainly be the case if the differentiated image has a less uniform distribution of gray levels (or spectral values) than the original image, as we would normally expect.

Match positions can be determined to within less than a pixel by resampling the image (at a grid of points that have fractional coordinates, in terms of the original sampling grid), and assigning gray levels to the new pixels by interpolation. The match at such an interpolated position may be better than the match at any of the original (intercoordinate) positions. The investigation of interpolated match positions is best done as a fine tuning of a coarse match position found without use of interpolation. The interpolated images will tend to have more uniform gray-level or spectral distributions than the original image, so that the matching speed-up process will probably not be as effective during this fine-tuning stage as at the coarse search stage.

A refinement allowing more accurate (part pixel) registration, when such accuracy is warranted, may be described as follows:

Let $F(x,y)$ be the cross-correlation function in terms of picture displacements (x,y) from the assumed best match point, $x = 0$, $y = 0$. Fit a quadratic to $F(x,y)$ of the form

$$F(x,y) = ax^2 + 2hxy + by^2 + 2gx + 2fy$$

We may determine the coefficients from the values of F at a few points; at $x = n$, $y = 0$, for example:

$$F(n,0) = an^2 + 2gn$$

Similarly,

$$F(-n,0) = an^2 - 2gn$$

hence

$$a = [F(n,0) + F(-n,0)]/2n^2$$
$$g = [F(n,0) - F(-n,0)]/4n$$

[2]It may be of further advantage to use a thresholded outline image so as to reduce the effects of overall gray-level differences due to seasonal changes, etc.

We can also find

$$b = [F(0,n) + F(0,-n)]/2n^2$$
$$f = [F(0,n) - F(0,-n)]/4n$$
$$h = [F(n,n) - F(n,0) - F(o,n)]/2n^2$$

Differentiating, $F(x,y)$ is a maximum when[3]

$$ax_0 + hy_0 + g = 0$$
$$\text{and } by_0 + hx_0 + f = 0$$

so that,

$$x_0 = (hf - bg)/(ab - h^2)$$
$$y_0 = (gh - af)/(ab - h^2)$$

Then shift origin to (x_0,y_0) and repeat with $n/2$ area size. Since only five values of F need to be obtained per iteration, the speed of such a procedure may prove to be adequate. As an example in one dimension, consider the following template:

3 1 9 1 3

and the image line gray values:

6 6 4 8 5 6 6

Take $F(x) = ax^2 + 2gx$.

$$F(-1) = \frac{(3 \times 6) + (1 \times 6) + (9 \times 4) + (1 \times 8) + (3 \times 5)}{\sqrt{(3^2 + 1^2 + 9^2 + 1^2 + 3^2)(6^2 + 6^2 + 4^2 + 8^2 + 5^2)}}$$

$$= .6208 = a - 2g$$

$$F(1) = \frac{(3 \times 4) + (1 \times 8) + (9 \times 5) + (1 \times 6) + (3 \times 6)}{\sqrt{(3^2 + 1^2 + 9^2 + 1^2 + 3^2)(4^2 + 8^2 + 5^2 + 6^2 + 6^2)}}$$

$$= .6656 = a + 2g$$

$$a = 1/2[F(1) + F(-1)] = .6432$$

$$g = 1/4[F(1) - F(-1)] = .0012$$

Differentiating, $F(x)$ is a maximum when $ax_0 + g = 0$, that is, for $x_0 = -g/a = -.0174$.

Shifting the origin to $x = -.0174$, we see

[3]After determining coefficients, we require that F be positive definite, that is, $ab > h^2$, to guarantee a true maximum of F.

Gray scale	Old coordinate	New coordinate
6	−3	−2.9826
6	−2	−1.9826
4	−1	−.9826
8	0	.0174
5	1	1.0174
6	2	2.0174
6	3	3.0174

In this new coordinate system we will need $F(-.5)$ and F(.5); hence, by interpolation, we will need the gray-level values at $x = -2.5, -1.5, -.5, .5, 1.5$, and 2.5. If linear interpolation is used, then

$$G(x) = G(x_1) + \frac{[G(x_2) - G(x_1)}{(x_2 - x_1)}(x - x_1)$$

so

$$G(-2.5) = 6.0000$$
$$G(-1.5) = 5.0348$$
$$G(-0.5) = 5.9304$$
$$G(.5) = 6.5522$$

etc. Then a new iteration is undertaken and the process continues until the origin shift indicated is satisfactorily small.

A possibility for still further improvement in matching efficiency is to identify locations in the image where the match can be expected to be good. This is typically done by making some simple local measurements on the image and finding positions where the values of these measurements are close to their values for the template. In the present application, the positional uncertainty of the landmarks is not expected to be very large, so that it seems unlikely that much will be gained by the use of this approach.

There is, however, an important possibility for reducing the number of match positions that should be tried once a match has been found for the first landmark. If this match is correct, the positional uncertainty of the remaining landmarks has now been reduced, since there are now fewer degrees of freedom. This is true even if the match is regarded as only approximate. Thus, after the first match is found, the next match can be searched for over a relatively smaller region; and after enough matches have been found, additional matches have very little positional uncertainty, so that we can search for them over very small regions. Of course, if the subsequent matches are not found in the expected places, they must be sought for over wider regions; and the original matches must be reevaluated. Thus, a feedback process can be used to zero-in on a best combination of match positions.

Additional information on template matching can be found in Barnea and Silverman (1972), Hall, Endlich, Wolf, and Brain (1972), Lam and Hoyt (1972), and Webber (1973).

Multispectral Classification

One of the most widely accepted applications of pattern recognition techniques to image processing has been the assignment of pixels to land use categories through multispectral classification. The process employs multiple images of the same scene, obtained from several spectral regions as input, and produces as output both displays or maps of scene content on a pixel by pixel basis and statistical summaries of scene content in tabular form. Multispectral classification has found a great deal of usage in agricultural assessment, forest management, petroleum and mineral exploration, land use mapping, urban planning, ecological monitoring, and water resources information collection.

Often the input imagery is obtained from satellite sensors such as the Landsat MSS or aircraft scanners. Landsat MSS data is ideally suited for this form of processing because it is quantified, easily obtained in abundance, standardized in format, acquired by a stable platform, and radiometrically calibrated. The four spectral bands are acquired in registration simultaneously. The four values presented to the computer to characterize any one pixel correspond to the amount of energy radiated, that is, reflected or emitted, from a ground area defined by the scanner's instantaneous field of view. Often these four radiance values are referred to as the pixel's spectral signature.

Dozens of processing algorithms have been developed to perform this classification. Perhaps the most generally accepted is the maximum likelihood method, developed by the Laboratory for Applications of Remote Sensing (LARS) at Purdue University and implemented in the LARSYS computer program (Landgrebe, 1979; Lindenlaub, 1972; Swain, 1972).

For any particular pixel, consider the four radiance values from the four spectral bands to constitute a measurement vector, $X = (X_1, X_2, X_3, X_4)$, where for convenience we specialize to the case of Landsat MSS data. Suppose we wish to assign each of the pixels in a scene to one or another of five categories of land cover, for example, water, forest, bare soil, wheat, and pasture. The signatures associated with these categories of land cover constitute five statistical populations, assumed to be normally distributed. The probability density, p_i of X when X is from the ith ($i = 1, \ldots, 5$) population is given, except for normalizing factors, by

$$p_i = \exp[-\tfrac{1}{2}(X - M_i)' \, V \, (X - M_i)]$$

where M_i is the mean of the ith population, V is the inverse of the covariance matrix, and the prime indicates transpose.

The quantity U_{ij} is then calculated for the 20 possible ij combinations for i not equal to j, where

$$U_{ij} = \log(p_i/p_j)$$

or more explicitly,

$$U_{ij} = X'V(M_i - M_j) - \tfrac{1}{2}(M_i + M_j)'V(M_i - M_j)$$

Then the program picks the largest of these and assigns the pixel to the category corresponding to the first subscript. For example, if the largest of the U_{ij} occurs for $i = 3, j = 4$, then the pixel is assigned to the category 3, in this case, bare soil. One assumption of this method not mentioned previously and often ignored in practice is that the pixels come with equal probability from any of the five (or more generally, m) populations being compared.

One must train the classifier to execute this algorithm by ascertaining values for the population means and the covariance matrix elements. This training is either supervised or unsupervised. In supervised training, the user presents the computer with the spectral signatures from a relatively small set of pixels he or she has identified as coming from each of the categories under consideration. The user does this on the basis of either manual interpretation or ground truth. Typically, this set of training pixels will constitute less than 1% of the full scene, whereupon the computer can classify rapidly and automatically the remaining 99% of the scene.

In unsupervised training, the user allows the computer to select the population means and covariance matrix elements by performing a clustering process on the input data. The user specifies the number of categories he or she seeks to discriminate among, and the computer selects that number of local density maxima in spectral signature space and then partitions the space accordingly. Later, the user can determine which type of land cover corresponds to each of the classes generated by the clustering. One popular clustering algorithm is termed ISODATA.

Both approaches to training can encounter situations in which two classes cannot be separated spectrally. For example, wheat and barley may have very similar spectral signatures in a particular Landsat frame. In this circumstance, it may be necessary to combine these two classes into one class termed *small grains*. Because of this limit on discriminability, an iterative approach involving both supervised and unsuperivsed training is often employed. One aspect of this is that, under unsupervised training, it is difficult to know how many classes are separable using spectral signatures. With iterative analysis, a variety of unsupervised clusterings can be tried. On the other hand, many classes of high interest to the user are differentiated only by subtle spectral differences, whereas some classes that are easily separable spectrally are of low user interest. Supervised training can prevent the needless discrimination of low-interest categories. Consequently, a relatively high level of man–machine interface is indicated for assessments of this type. These procedures have made available detailed levels of land cover status information determination previously unavailable in such a timely fashion, (because it is so geographically dispersed), and at such a low cost. This has given rise to more effective resource management methods and, consequently, to improved productivity.

The spectral characteristics of various types of vegetation change drastically as a function of time. The changes occur both in the course of a growing season and from one year to the next. As a crop matures, the plants change their spectral characteristics continuously. Early in the growing season, for example, the instantaneous field of view of the scanner may sense, primarily, the spectral signature of bare soil, and hence the signature characteristic of the crop is relatively suppressed. Furthermore, on the same calendar date from one year to the next, the crops are likely to be at different levels of maturity since weather conditions determine the planting date for most species. This implies that the leaf area, or canopy, to soil area ratio will change on a given date from year to year. This applies as well to the changes in the spectral signature of a crop in the course of a year, for example, corn tasseling.

Just as there are limits to the applicability of spectral signature information that is collected at one time and used to classify data collected at another time, there are also limits on the spatial extent to which spectral signatures can be applied. Changes in soil type, climate, elevation, and other factors can cause a crop to exhibit spectral signature differences even in relatively nearby geographic areas. And these sources of variation are compounded by the effects of atmospheric attenuation, which vary unpredictably in both space and time.

The LARSYS program consists of four sections. These are (*a*) statistical analysis, (*b*) feature selection, (*c*) pattern classification, and (*d*) results display. The statistics processor produces histograms of the spectral signatures of all the pixels the user has designated in the training set for each class, tables of the means and standard deviations for each class, the correlation matrices, multispectral response graphs, and other statistical descriptors of the training data. These analyses can provide guidance to the user regarding the prospects for a successful classification. With some experience, for example, an analyst will recognize that an abnormally large standard deviation signifies that pixels of differing land cover types have been intermixed in a single class. Similarly, he or she may observe that two classes have the same multispectral response curves and are, in fact, the same class.

Some multispectral scanners produce 12, 16, or even more spectral channels of imagery. Processing all of this data requires a substantial amount of computer time and hence can be fairly expensive. Frequently, very little improvement in classification accuracy is obtained from using, say, 16 channels rather than 3 or 4. But selecting the optimal 3 or 4 from the hundreds of possible combinations requires the feature selection segment of the LARSYS program. This algorithm measures the separability between all pairwise combinations of classification categories for all possible combinations of spectral channels. Then it indicates the best choice.

The pattern classification processor segment of the LARSYS program applies the maximum likelihood algorithm described earlier to assign each pixel to one of the specified categories. In some cases, a likelihood threshold is employed to ensure that each classified pixel differs from the mean of the most likely class by no more than a specified number of standard deviations. Pixels that fail this test remain unclassified and possibly represent a class for which the algorithm was not trained.

One variation on the maximum likelihood classifier worthy of note is the Bayesian classifier. This algorithm is identical to the maximum likelihood classifier except that it does not assume that each of the categories are equally probable. If a user has a priori knowledge that one class is more probable than another, say forest is more probable than wheat, then using a Bayesian classifier can exploit this knowledge and improve classification accuracy.

The display segment of the LARSYS program reports the results in two formats. The first is an alphanumeric line printer map. Each classified pixel is represented by a user-specified symbol, for example, *C* for corn, *S* for soybeans, and *W* for water.

The other format for reporting results is tabular and involves the concept of test data. Just as a small set of pixels of known land cover type is used to train the classifier, so also is a small set of known pixels used to evaluate the accuracy of the classification. The tabular report of the accuracy for each category using this test data advises the analyst whether the training data was typical of the rest of the scene. Note that the classifier assigns each pixel of the training data to some category as well. If the classifier assigns a high proportion of the training pixels to categories different from their designation, then this could indicate a designation error or an algorithm assumption violation.

Naturally, the separability of land cover types depends in part on the time of the year the imagery is acquired. In mid-July, for example, it is very difficult to discriminate between corn and soybeans. At that time both present the scanner with a view of a solid green canopy. In early August, however, corn tassels appear and discrimination becomes much more reliable.

Unfortunately, the time for optimal discrimination does not occur simultaneously among all class pairs. This difficulty can be overcome by combining data from scenes acquired at different times. Landsat frames obtained on two different dates but of the same geographic area can be registered and the eight spectral channels input to the classifier program as if they were eight channels of the same scene. These multispectral–multitemporal signatures are often more separable than are single date multispectral signatures. For example, in the early fall, bare soil is likely to indicate fallow fields or fields being prepared for planting wheat. In the early spring, wheat fields are lush green, but so are pastures. By using data from both early fall and early spring, wheat can be discriminated from both pastures and fallow fields. Unfortunately, there are often small misregistration errors in overlaying imagery in this fashion. If the analysis involves scenes with fine ground detail, then these misregistration errors can degrade the classification accuracy more than the multitemporal dimension of the data can enhance it.

In the same way that both the spectral dimension and the temporal dimension allow pattern recognition methods to classify pixels into land cover categories, the spatial dimension can be used for this process, too. Texture classification alone is generally not as accurate as multispectral classification, but combined they are often more accurate than either is alone. This approach has been explored at LARS with the ECHO classifier (Landgrebe, 1978).

Eigenpictures

Multispectral classification, described qualitatively in the previous section, can be described mathematically in terms of vectors. If we are presented with a three-band multispectral image, we may represent each pixel of that image as a point in three-dimensional space. The coordinates of a given pixel in that space are the gray levels in the three bands of that pixel. Said differently, each pixel can be represented as a vector; the first component of that vector is the gray level of that pixel in the first spectral band, the second component is the gray level of that same pixel in the next band, etc. The coordinate axes are labeled *band 1, band 2,* and *band 3.* In the case of Landsat MSS, of course, we must deal with a four-dimensional space, one for each band, but in this discussion we will consider a three-band image to help visualization.

Multispectral classification is the process of partitioning that space into regions. If the image is an agricultural scene and the partitioning is done well, each of the spatial regions will correspond to a particular type of crop, say wheat or barley. There are many methods for partitioning the space into regions for multispectral classification. The LARSYS method described previously is a statistically based method termed *maximum likelihood,* a special case of a more general method termed *Bayesian classification.* Another method, termed *parallelopiped,* partitions the space using planes parallel to the axes. Still a different method assigns a pixel to this region or that on the basis of the Euclidean distance of that pixel's coordinate point from the region's centroid. The method selected for a particular task depends on what assumptions you are prepared to make about the data set under analysis.

This vector formalism can be used for ends other than multispectral classification. One such use is the generation of an eigenpicture. An *eigenpicture* is a color composite picture of the original multispectral image data with colors mixed in a way that optimizes the visual discrimination among differently colored areas. One may think of it as adjusting the hue and tint of a color television to obtain the most detailed picture.

Mathematically, the eigenpicture is obtained by rotating the axes of the representation space to maximize the variance of the data in the dimensions that are to be used to produce the composite color picture. The gray levels of each pixel in the new, that is, rotated, coordinate system are obtained as linear combinations of the original gray levels for that pixel. On occasion, interpretation of an eigenpicture is dramatically improved over that of the standard color composite image. (See the discussion of principal components in Pratt, 1978 and in Rosenfeld and Kak, 1976.)

EDGE AND LINE DETECTION

Much information about an image is conveyed by its lines and edges. Determining the paths of roads, rivers, railroads, and geologic faults constitutes a major activity within the field of digital image processing. Ascertaining the edges of agricultural fields, lakes, forests, floodplains, snow cover, and other terrain regions is a closely allied activity that also merits treatment in a survey of this type. An example of some detail showing the use of region boundary detection is given in Chapter 6.

Edge enhancement, treated on pages 78–79, consists of processing images to generate an image with sharper edges for improved interpretability. Edge detection, on the other hand, is a type of image analysis wherein a report in the form of an edge map is extracted from an input image. A thorough survey of edge detection techniques can be found in Davis (1975). Also of interest are Nevatia (1977), Pratt (1978), Schachter, Lev, Zucker, and Rosenfeld (1977), Shaw (1979), and Wechster and Kidode (1977).

Gradient

The gradient image is obtained by subtracting the gray-level value of each pixel from the gray-level value of a neighboring pixel. For example, if the neighboring pixel is chosen from the same image row, a directional derivative in the X direction is obtained; if the neighboring pixel is chosen from the same image column, a directional derivative in the Y direction results and can be represented as dG/dy. The isotropic first derivative, or gradient, is computed

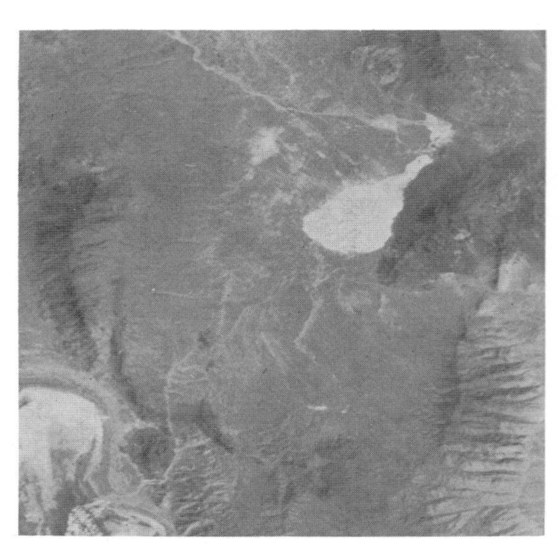

FIGURE 4.16. A window from band 5 of a Landsat MSS scene over Nevada.

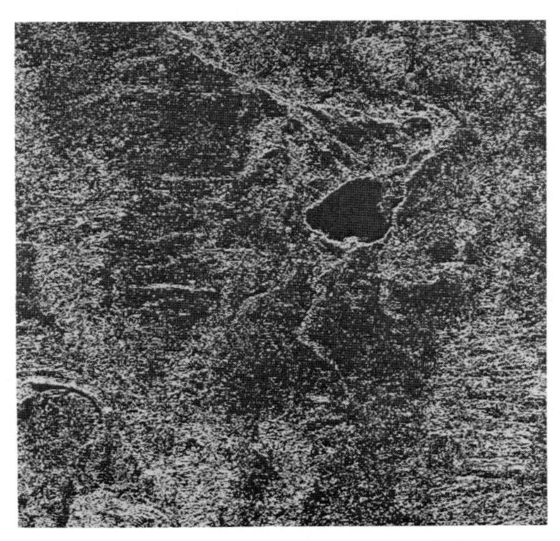

FIGURE 4.17. The directional derivative in the Y direction, which emphasizes edges in the X direction.

as the square root of the sum of the squares of those two directional derivatives:

$$\text{grad } G = [(dG/dx)^2 + (dG/dy)^2]^{\frac{1}{2}}$$

High values for grad G indicate the presence of edges or abrupt changes in gray-level value. Other forms of the gradient are in common use, as, for example, in Chapter 6 (page 208). Figure 4.16 shows a window from band 5 of a Landsat MSS scene over Nevada. Figure 4.17 shows the directional derivative in the Y direction (which emphasizes edges in the X direction), and the X directional derivative is shown in Figure 4.18. Generally the gradient is preferred for finding lines, whereas the Laplacian (see page 78) is more useful in edge and boundary detection. The Laplacian is a second derivative and is consequently more troubled by image noise.

The Hough Transform[4]

In 1972, Duda and Hart reported the use of the Hough transformation to detect lines and curves in binary pictures. They suggested an alternative

[4]This section originally appeared in R. Michael Hord, Extending the Hough Transform, in the *Proceedings of the Seventh Annual Automatic Pattern Recognition Symposium*, Electronic Industries Association, Washington, D.C., 1977. Reprinted with permission.

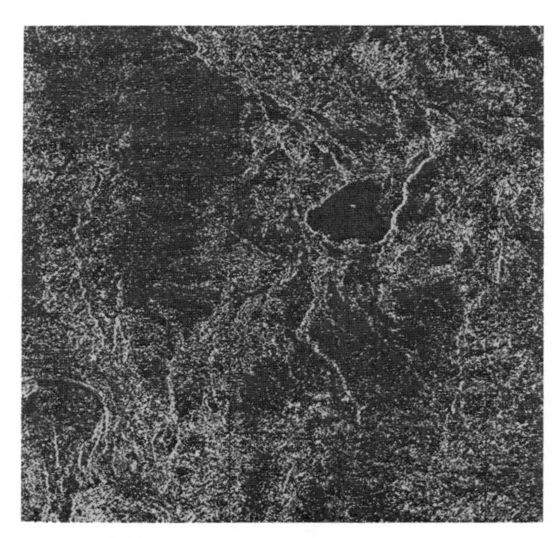

FIGURE 4.18. The X directional derivative.

parameterization to that used by Hough (1962) 10 years previously, as reported by Rosenfeld (1969). Briefly, the Hough transform replaces the original problem of finding colinear points by a mathematically equivalent problem of finding concurrent lines (Pratt, 1978). Suppose we have a set of N points (x_i, y_i), $i = 1, 2, \ldots, N$. The lines through a given one of these points (x_i, y_i) are given by

$$P = x_i \cos \theta + y_i \sin \theta \qquad (5)$$

where θ specifies the angle of the normal to the line with respect to the x axis and P is the distance of the line from the origin, as shown in Figure 4.19. Note that the range of P is both positive and negative, but bounded by the format size. Concurrent lines are characterized by the same P, θ parameter values. Quantizing P and θ into suitable increments allows the generation of a two-dimensional histogram of all of the P, θ values associated with the N points; for each (x_i, y_i) and for each θ increment, Eq. 5 indicates a specific P value. If the accumulated count of P, θ occurrences, the histogram, is termed $H(P, \theta)$, then high values of H indicate a large number of co-linear points. An example of H (P, θ) is shown in Figure 4.20. The left half of this H display reports line counts with P ranging from 0 (top) to 63 (bottom), with θ ranging from 6° (left) to 180° (right); the right half shows counts for P between -1 (top) and -64 (bottom). The 10 lines correspond to the 10 asterisks in H.

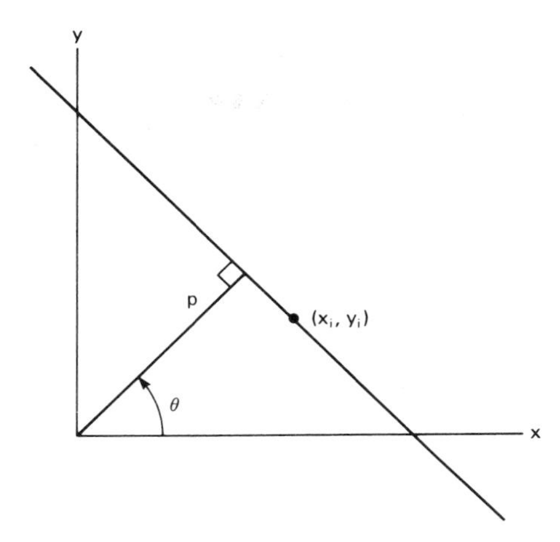

FIGURE 4.19. Hough transform geometry. Theta specifies the angle of the normal to the line with respect to the x axis; P is the distance of the line from the origin.

In practice, then, a digital image can be subjected to a derivative and threshold operation to generate a set of N candidate line and edge elements. Then the Hough histogram is generated and thresholded. The surviving P, θ pairs indicate image lines that can be used to eliminate spurious line and edge candidate elements and to fill in gaps of the detected lines and edges.

As Duda and Hart (1972) point out, a similar method can be used to find just the lines through a given point, just the lines of a given direction, and, by extension to higher dimensions for H, any arbitrary type curve. Shape detection (see Chapter 6) and template matching (See page 88) are topics unto themselves. The Hough transform can be of some use in these areas, however.

ANGLE AND SHAPE DETECTION

For detection of shapes with straight line edges figures of merit based on combinations of $H\ (P,\theta)$ values can easily be designed. For example, suppose we seek triangular shapes, arbitrary in size, orientation, and shape. We note that when two lines meet at an angle α, then the normals to those lines intersect at the same angle α, as shown in Figure 4.21. Hence the property that the sum of the angles of any triangle is 180° can be translated to a relationship among $H(P,\theta)$ values.

(a)

```
00001000010000100001000010000100
00001000010000100001000010000100
00001000010000100001000010000100
00001000010000100001000010000100
00001000010000100001000010000100
11111111111111111111111111111111
00001000010000100001000010000100
00001000010000100001000010000100
00001000010000100001000010000100
00001000010000100001000010000100
00001000010000100001000010000100
00001000010000100001000010000100
00001000010000100001000010000100
00001000010000100001000010000100
00001000010000100001000010000100
00001000010000100001000010000100
00001000010000100001000010000100
00001000010000100001000010000100
11111111111111111111111111111111
00001000010000100001000010000100
00001000010000100001000010000100
00001000010000100001000010000100
00001000010000100001000010000100
00001000010000100001000010000100
00001000010000100001000010000100
11111111111111111111111111111111
00001000010000100001000010000100
00001000010000100001000010000100
00001000010000100001000010000100
00001000010000100001000010000100
00001000010000100001000010000100
11111111111111111111111111111111
```

(b)

```
000000000000001232233432322000 000000000000000012323443333221
000000000000011432233332221000 000000000000000012123344332231
100000000000011422333332111000 000000000000000012222344333341
321001000100111432323322110000 00000000000000000112233223333★
42100000111224★111222222210000 000000000000000000212233233211
321100010011341112234321110000 000000000000000000111234432221
221101111112341112333322100000 000000000000000000111223333231
132211111122321113232221100000 000000000000000000011222332441
433110111112311112333211100000 00000000000000000001112332333★
333211111222311122333220000000 000000000000000000011123333211
432122111123211132332111000000 000000000000000000001112333221
123222221222111122333211000000 000000000000000000000111232231
133322222233211133332100000000 000000000000000000001112333441
433321222222111333232110000000 00000000000000000000011223333★
332333222322211433421110000000 000000000000000000000011233211
423332223321111434211100000000 000000000000000000000011233221
123333333332221432321100000000 000000000000000000000011132231
13334232232234★122221000000000 000000000000000000000001123441
332324333222341133221000000000 00000000000000000000000111333★
433323343222341133210000000000 000000000000000000000000113211
323244323322321333210000000000 000000000000000000000000112321
123432444323311433100000000000 000000000000000000000000012331
132334443332211432100000000000 000000000000000000000000002341
333333333333321431100000000000 00000000000000000000000001133★
43333434323334★121000000000000 000000000000000000000000001111
423332333333341121000000000000 000000000000000000000000011221
123333344332341310000000000000 000000000000000000000000001231
133333322433221410000000000000 000000000000000000000000000141
432323333342311310000000000000 00000000000000000000000000002★
323221233324421200000000000000 000000000000000000000000000001
42323222223334★000000000000000 000000000000000000000000000001
122223223223230000000000000000 000000000000000000000000000000
021212222222230000000000000000 000000000000000000000000000000
022220221222200000000000000000 000000000000000000000000000000
001122112221100000000000000000 000000000000000000000000000000
001111221111100000000000000000 000000000000000000000000000000
000111111111000000000000000000 000000000000000000000000000000
000110111100000000000000000000 000000000000000000000000000000
000011010100000000000000000000 000000000000000000000000000000
000001101000000000000000000000 000000000000000000000000000000
000000000000000000000000000000 000000000000000000000000000000
000000000000000000000000000000 000000000000000000000000000000
000000000000000000000000000000 000000000000000000000000000000
000000000000000000000000000000 000000000000000000000000000000
000000000000000000000000000000 000000000000000000000000000000
000000000000000000000000000000 000000000000000000000000000000
000000000000000000000000000000 000000000000000000000000000000
000000000000000000000000000000 000000000000000000000000000000
000000000000000000000000000000 000000000000000000000000000000
000000000000000000000000000000 000000000000000000000000000000
000000000000000000000000000000 000000000000000000000000000000
000000000000000000000000000000 000000000000000000000000000000
000000000000000000000000000000 000000000000000000000000000000
000000000000000000000000000000 000000000000000000000000000000
000000000000000000000000000000 000000000000000000000000000000
000000000000000000000000000000 000000000000000000000000000000
000000000000000000000000000000 000000000000000000000000000000
000000000000000000000000000000 000000000000000000000000000000
000000000000000000000000000000 000000000000000000000000000000
000000000000000000000000000000 000000000000000000000000000000
000000000000000000000000000000 000000000000000000000000000000
000000000000000000000000000000 000000000000000000000000000000
000000000000000000000000000000 000000000000000000000000000000
```

FIGURE 4.20 (*a* and *b*). An example of a grid pattern output.

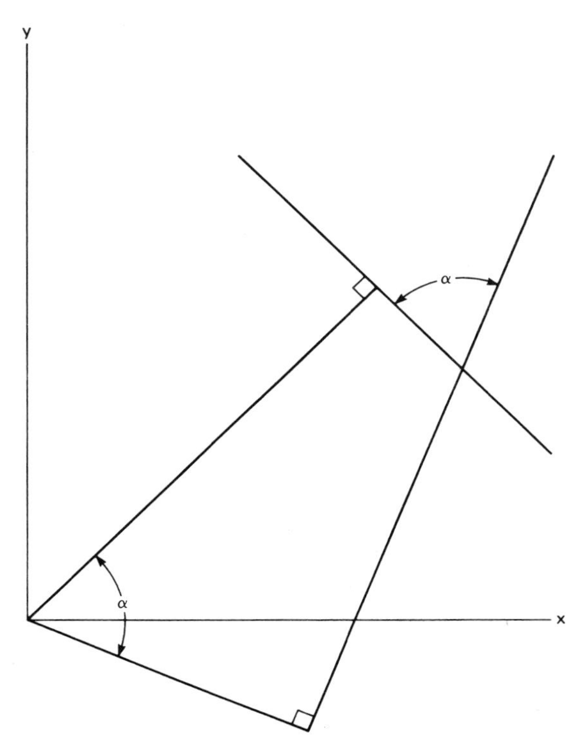

FIGURE 4.21. Angle detection geometry. When two lines meet at an angle, α, then the normals to those lines intersect at the same angle.

Define

$$G(\theta) = [\max H(P,\theta);\ P_{min} < P < P_{max}] \quad (6)$$

$$T(\theta_1, \theta_2) = G(|\theta_1 - \theta_2|)G(|\theta_2 - \theta_3|)G(|\theta_3 - \theta_1|) \quad (7)$$

where

$$|\theta_1 - \theta_2| + |\theta_2 - \theta_3| + |\theta_3 - \theta_1| = 180° \quad (8)$$

$T(\theta_1, \theta_2)$ values that exceed a threshold indicate a triangle detection. Naturally, if a specific triangular shape is sought, the search calculations are greatly reduced. Figure 4.22a shows an example. In Figure 4.22b, three peaks are shown:

$$\theta_1 = 90°$$
$$\theta_2 = 138°$$
$$\theta_3 = 180°$$
$$\theta_1 - \theta_2 = 48°$$
$$\theta_2 - \theta_3 = 42°$$
$$\theta_3 - \theta_1 = 90°$$
$$|\theta_1 - \theta_2| + |\theta_2 - \theta_3| + |\theta_3 - \theta_1| = 180°$$

TEMPLATE ORIENTATION

Often one seeks to locate a large format image where a small format template can be placed congruently. If the template pattern is rotated with

```
    000000000000000000000000000000
    010000000000000000000000000000
    011000000000000000000000000000
    010100000000000000000000000000
    010010000000000000000000000000
    010001000000000000000000000000
    010000100000000000000000000000
    010000010000000000000000000000
    010000001000000000000000000000
    010000000100000000000000000000
    010000000010000000000000000000
    010000000001000000000000000000
    010000000000100000000000000000
    010000000000010000000000000000
    010000000000001000000000000000
(a) 011111111111111100000000000000
```

```
    000000000000000001224322232100  000000000000000000000028*4446840
    521000000000000111224112222000  00000000000000000000000002422226*
    532211000011111122222222220000  000000000000000000000000003221111
    142121211100111111222122210000  000000000000000000000000002222211
    122222121111111111223222120000  000000000000000000000000000321111
    102222102111011112222222200000  000000000000000000000000000222111
    111210110111111112222222100000  000000000000000000000000000021211
    111022221101111212222211000000  000000000000000000000000000022111
    110122111120111111221200000000  000000000000000000000000000002121
    111110111111111112222200000000  000000000000000000000000000001111
    011202111011101112222100000000  000000000000000000000000000000211
    101011211102011132210000000000  000000000000000000000000000000111
    111111011211111143100000000000  000000000000000000000000000000011
    110110111010111631000000000000  000000000000000000000000000000011
    111112222222111141000000000000  000000000000000000000000000000001
    111011111121246*00000000000000  000000000000000000000000000000000
    021110011223440000000000000000  000000000000000000000000000000000
    001101212122300000000000000000  000000000000000000000000000000000
    000112122222000000000000000000  000000000000000000000000000000000
    000021001210000000000000000000  000000000000000000000000000000000
    000000221100000000000000000000  000000000000000000000000000000000
    000000000000000000000000000000  000000000000000000000000000000000
(b) 000000000000000000000000000000  000000000000000000000000000000000
```

FIGURE 4.22. (*a*) An example of triangle input. (*b*) The Hough transform of the triangle.

respect to the corresponding pattern in the search image, finding the match can be a problem. If, however, the template pattern exhibits lines, for example, fiducial marks, then the Hough transform may be used. One transforms the template and successively transforms windows from the search image. Then template matching in the transform space consists of the translational search along the (P,θ) axes that would have been performed in image space (x,y) axes had the template not been rotated. On finding a match, the θ offset in Hough space indicates the angle through which the template should be rotated in image space to achieve congruence.

TEMPLATE SCALING

Similarly, for a template represented at a different scale from that of the search image, finding the proper template scale adjustment in image space can be difficult. In Hough transform space, however, a characteristic scale can be obtained by calculating the distance between the two highest histogram peaks. The characteristic scale of the search windows is then the ratio of these characteristic scales. (See Figures 4.23 and 4.24.)

```
                            000000000000000000001001000000  000000000000000000012222121110
                            000000000000000000101110100000  000000000000000000000100000000
                            000000000000000000011010100000  000000000000000000000111011000
                            000000000000000000111111000000  000000000000000000000111100000
                            000000000000000001011000000000  000000000000000000000000000000
                            000000000000000001211111000000  000000000000000000000010000000
0000000000000000000000000   000000000000000011210010000000  000000000000000000000011000000
0000000000000000000000000   000100000000000021111100000000  000000000000000000000000000100
0000000000000000000000000   000010000000000012211000000000  000000000000000000000001000000
0000000000000000000000000   000100000000000022211000000000  000000000000000000000000100000
0000000000000000000000000   000000000000000422200000000000  000000000000000000000000110100
0000000000000000000000000   000000000000000321000000000000  000000000000000000000000000000
0000000000000000000000000   00000110001124*021000000000000  000000000000000000000000010000
0000000000000000000000000   000001010001130021100000000000  000000000000000000000000000000
0000000000000000000000000   000010111011220221000000000000  000000000000000000000000000000
0000000000000000000000000   001010111111200421000000000000  000000000000000000000000001000
0000000000000000000000000   000011101222003000000000000000  000000000000000000000000000000
0000000000000000000000000   0001010111222 4*000000000000000  000000000000000000000000000000
1111111111111111111111111   000000111112240000000000000000  000000000000000000000000000000
0000000000000000000000000   000000110121220000000000000000  000000000000000000000000000010
0000000000000000000000000   000111011111200000000000000000  000000000000000000000000000000
0000000000000000000000000   000011111111000000000000000000  000000000000000000000000000000
0000000000000000000000000   001001111101000000000000000000  000000000000000000000000000000
1111111111111111111111111   000000110110000000000000000000  000000000000000000000000000000
0000000000000000000000000   000011110000000000000000000000  000000000000000000000000000000
0000000000000000000000000   000001100000000000000000000000  000000000000000000000000000000
0000000000000000000000000   000010000000000000000000000000  000000000000000000000000000000
0000000000000000000000000   000010000000000000000000000000  000000000000000000000000000000
0000000000000000000000000   000000000000000000000000000000  000000000000000000000000000000
0000000000000000000000000   000000000000000000000000000000  000000000000000000000000000000
0000000000000000000000000   000000000000000000000000000000  000000000000000000000000000000
```

FIGURE 4.23. A template scale adjustment obtained in Hough transform space, peaks 5 units apart.

```
000000000000000000000000
000000000000000000000000
000000000000000000000000
000000000000000000000000
000000000000000000000000
000000000000000000000000
000000000000000000000000
111111111111111111111111
000000000000000000000000
000000000000000000000000
000000000000000000000000
000000000000000000000000
000000000000000000000000
000000000000000000000000
000000000000000000000000
000000000000000000000000
000000000000000000000000
111111111111111111111111
000000000000000000000000
000000000000000000000000
000000000000000000000000
000000000000000000000000
000000000000000000000000
000000000000000000000000
000000000000000000000000
```

```
000000000000000001001100000000
000000000000000001101100100000
000000000000000021011111000000
000000000000000021111100000000
000000000000000221011000000000
000000000000000321210000000000
000000000000000310100000000000
00000000001123*000000000000000
000000000001240000100000000000
000110000001122000100000000000
001001000001200001100000000000
000000100001100011000000000000
000001000001100002100000000000
000000010001000021100000000000
000010111010000221000000000000
001111011111000042100000000000
000001111110000300000000000000
00000001111124*000000000000000
000000011101240000000000000000
000001110011220000000000000000
000111001001200000000000000000
000000010011100000000000000000
000010100001000000000000000000
000001000011000000000000000000
000000000000000000000000000000
000000000000000000000000000000
000000000000000000000000000000
```

```
000000000000000001212222121110
000000000000000000000000000000
000000000000000000100110101000
000000000000000000000001100000
000000000000000000000000010000
000000000000000000000011000000
000000000000000000000000000000
000000000000000000000000001100
000000000000000000000001100000
000000000000000000000001000000
000000000000000000000000110000
000000000000000000000000000000
000000000000000000000000000000
000000000000000000000000000100
000000000000000000000000000000
000000000000000000000000010000
000000000000000000000000000000
000000000000000000000000000000
000000000000000000000000000000
000000000000000000000000000010
000000000000000000000000000000
000000000000000000000000000000
000000000000000000000000000000
000000000000000000000000000000
000000000000000000000000000000
000000000000000000000000000000
000000000000000000000000000000
```

FIGURE 4.24. A template scale adjustment obtained in Hough transform space, peaks 10 units apart.

LINE FOLLOWING

Many line-following algorithms are in common use. Each produces lines characterized by the constraints in the algorithm governing the selection of the "next point." An example of such a constraint is a local limitation on line curvature. Using the Hough transform, one can select a sequence of "next" points that all satisfy the global curvature restrictions imposed by the angular resolution selected in the transform.

LINE STRAIGHTENING

In the course of an image processing sequence, crooked lines occasionally emerge even in circumstances for which there is a priori information that these lines are straight. For example, this can occur as the result of skeletonizing (medial axial transform) (see page 113). The Hough transform, using low-resolution increments in P and θ, can find these lines; then the line elements input to the transform can be replaced by the straight line specified by the indicated (P,θ) pair.

WEIGHTED HOUGH TRANSFORMS

Several weighting schemes seem interesting. Obviously, one need not apply the Hough transformation to thresholded (binary) derivative images. If it is applied to the derivative image directly so that $H(P,\theta)$ cells contain not simply the count of co-linear elements, but rather the sum of co-linear derivative values, then strong short lines and strong but gappy lines can be detected as well as longer, weaker lines.

Then, too, the Hough transform from one spectral band of a multispectral image can be divided by the transform from another spectral band. This emphasizes lines and edges that are spectrally dependent. Applications for this include the attempt to find soil type boundaries from plant signature changes and camouflage detection.

CONCLUSION

The above enumeration is by no means exhaustive, but the thrust should be clear. The Hough transform is an efficient, effective detector on lines since the computational effort grows linearly with N rather than as N^2 for considering all pairs of figure points. Since lines and edges occur in most digital images and since these lines and edges convey so much of the image content, it is fortunate that such an efficient detector is available. This is particularly true in light of the growing importance of parallel processors since the Hough transform exhibits substantial parallelism.

Skeletonizing

The general problem of image pattern recognition is to assign each of the patterns to be recognized to one of a prescribed number of classes. The specific class to which a given pattern is assigned is chosen on the basis of the values assumed by certain measurements applied to the pattern. These measurements are termed *features*. The effectiveness of a pattern recognition process depends largely on the significance of the features.

Image pattern recognition is often associated with multispectral classification, but it is also concerned with other types of patterns, for example, spatial and temporal patterns. In the spatial domain, patterns can be assigned to classes on the basis of size, shape, location, or orientation. If one seeks to recognize patterns on the basis of size, then features are chosen that are

P9	P8	P7
P2	P1	P6
P3	P4	P5

FIGURE 4.25. Pixel numbering of picture elements.

independent of shape, location, and orientation. Such features are then termed *invariant* with respect to these other characteristics.

Generally, invariant features cannot be obtained directly from the original picture. Rather, the picture is subjected to some preprocessing that presents the information in a transformed way. This transformation process is termed *feature extraction*. One example of feature extraction is the two-dimensional Fourier transform, which is invariant with respect to the location of the pattern in the format of the original image.

If, in binary images, elongated objects of varying thickness are to be classified without regard to thickness, then a transform that makes the features' thickness invariant would be useful. A number of such transforms are

```
0 0 0 0 0 0 0 0 0 0 0 0 0 0 0 0
0 0 0 0 0 0 0 * * * 0 0 0 0 0 0
0 0 0 0 0 0 0 * * * * 0 0 0 0 0
0 0 0 0 0 0 * * * * * * 0 0 0 0
0 0 0 0 0 * * * * * * * 0 0 0 0
0 0 0 0 * * * * 0 0 * * 0 0 0 0
0 0 0 0 * * * * 0 0 * * 0 0 0 0
0 0 0 0 * * * 0 0 0 0 * * 0 0 0
0 0 0 * * * * * * 0 0 * * 0 0 0
0 0 * * * * * * * * * * * 0 0 0
0 0 * * * * * * * * * * * * 0 0
0 0 * * * 0 0 0 0 0 0 0 * * 0 0
0 0 * * * 0 0 0 0 0 0 0 * * 0 0
0 * * * 0 0 0 0 0 0 0 0 * * * 0
0 * * 0 0 0 0 0 0 0 0 0 0 * * 0
0 0 0 0 0 0 0 0 0 0 0 0 0 0 0 0
```

FIGURE 4.26. Input image, ones shown as asterisks.

described in the image pattern recognition literature. They are referred to as *thinning, skeletonizing* (Pratt, 1978) or medial *axial transforms* (Arcelli & Baja, 1980). We will focus on one reported by Stefanelli and Rosenfeld (1971). Skeletonizing is also used in Chapter 6 (page 210).

If we refer to a given picture element as $P1$, then we define $P2, \ldots, P9$, the neighbors of $P1$, as shown in Figure 4.25. Let $A(P1)$ be the number of 01 patterns in the ordered set $P2, P3, \ldots, P9, P2$. Let $B(P1)$ be the number of nonzero neighbors of $P1$. Then a nonzero point $P1$ is changed to zero in the image if all of the following four conditions prevail:

1. $2 \leq B(P1) \leq 6$
2. $A(P1) = 1$
3. $P2 * P4 * P8 = 0$ or $A(P2) \neq 1$
4. $P2 * P4 * P6 = 0$ or $A(P4) \neq 1$

This algorithm is applied iteratively until no further changes occur. This processing yields connected thinned figures. All the points of a figure can be processed simultaneously by a parallel processor. A small test case is shown in Figures 4.26 through 4.29.

```
0 0 0 0 0 0 0 0 0 0 0 0 0 0 0 0
0 0 0 0 0 0 0 0 0 0 0 0 0 0 0 0
0 0 0 0 0 0 0 0 * * 0 0 0 0 0 0
0 0 0 0 0 0 0 * * * * 0 0 0 0 0
0 0 0 0 0 0 * * 0 0 * 0 0 0 0 0
0 0 0 0 0 * * 0 0 0 * 0 0 0 0 0
0 0 0 0 0 * * 0 0 0 0 * 0 0 0 0
0 0 0 0 0 * * 0 0 0 0 * 0 0 0 0
0 0 0 0 * * * 0 0 0 0 * 0 0 0 0
0 0 0 * * * * * * 0 0 * 0 0 0 0
0 0 0 * * 0 0 0 * * * * * 0 0 0
0 0 0 * 0 0 0 0 0 0 0 0 * 0 0 0
0 0 * * 0 0 0 0 0 0 0 0 * 0 0 0
0 0 * 0 0 0 0 0 0 0 0 0 0 * 0 0
0 0 0 0 0 0 0 0 0 0 0 0 0 0 0 0
0 0 0 0 0 0 0 0 0 0 0 0 0 0 0 0
```

FIGURE 4.27. Picture after one iteration, 45 changes.

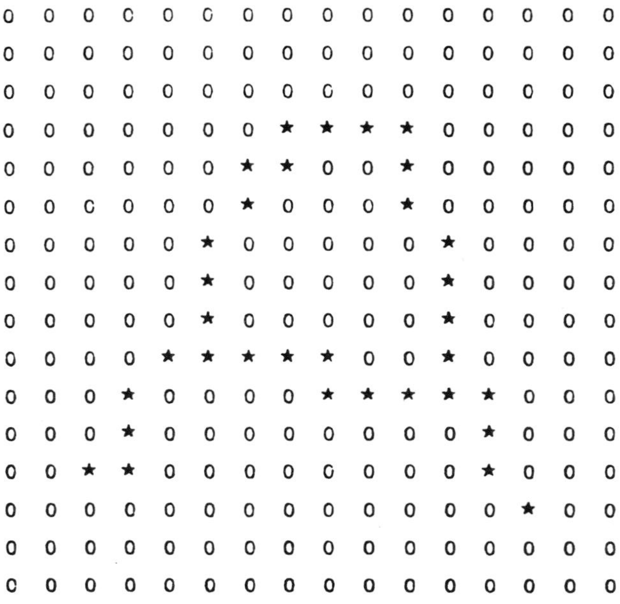

FIGURE 4.28. Picture after two iterations, 10 more changes.

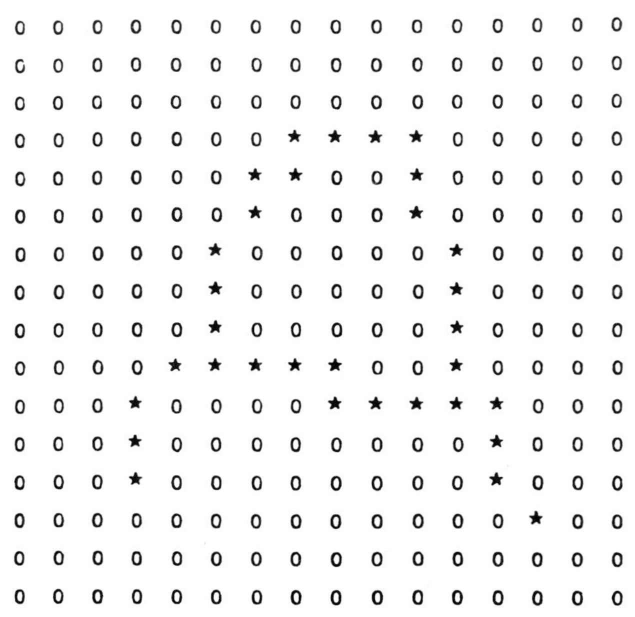

FIGURE 4.29. Picture after third and final iteration, one additional change.

OTHER OPERATIONS

Data Compression

The sheer volume of data in a digital image can give rise to practical problems. The transmission of digital imagery from one facility to another (Pratt, 1979) and the maintenance of an archive of digital imagery at a processing location can become fairly expensive as a result of the data-intensive nature of pictures represented as bits. Fortunately, this characteristic can be alleviated by data compression.

Image data compression is generally considered to be of two types, depending on whether or not the compression process preserves all of the information in the original, that is, on whether or not there is a decompression process that can retrieve the original image. The image reduction schemes described earlier constitute one type of noninformation preserving compression. Truncating the Fourier or Hadamard transform of an image constitutes another example. Examples of information preserving compression methods are also abundant. Run length codes and Differential Pulse Code Modulation are prominent techniques (Hung, 1979). Huffman coding is another compression process that permits recovery of the original image (Gallager, 1978).

Sample problem illustrating the use of Huffman coding to achieve minimum average length pixel codes:

The Southern California Landsat MSS frame ID 1094–18231 band 5 provided the data for this sample. Rows 500–2000 were scanned in increments of 10 rows, and columns 1000–2500 were scanned in increments of 10 columns to generate the 22650 element histogram in Table 4.2. The graph of this is shown in Table 4.3, where each asterisk represents 22.525 pixels. Deleting zero-entry cells provides the input histogram (Table 4.4) to the Huffman coding routine, which first rank sorts the histogram (Table 4.5). Huffman coding assigns a variable bit length code to each gray level, with more frequently occurring gray levels having shorter bit lengths. A prefix condition is imposed to ensure decodability. Table 4.6 shows the Huffman code for the sample data. Table 4.7 reports the bits per pixel by gray level and the average number of bits per pixel (4.09 bits per pixel). This compares with the 8 bits per pixel currently employed, a 49% savings.

DECODING

If, using the Huffman code of Table 4.6, a series of bits reading 01000101111100000 . . . is observed at the beginning of a record, we can decode into gray-level values as follows:

Table 4.2
Histogram of Landsat Window from Scene 1094–18231 Band 5

G(0) =	0	G(32) =	26	G(64) =	0	G(96) =	0	G(128) =	0	G(160) =	0	G(192) =	0	G(224) =	0
G(1) =	0	G(33) =	36	G(65) =	3	G(97) =	0	G(129) =	0	G(161) =	0	G(193) =	0	G(225) =	0
G(3) =	0	G(34) =	10	G(66) =	1	G(98) =	0	G(130) =	0	G(162) =	0	G(194) =	0	G(226) =	0
G(3) =	0	G(35) =	18	G(67) =	0	G(99) =	0	G(131) =	0	G(163) =	0	G(195) =	0	G(227) =	0
G(4) =	1	G(36) =	24	G(68) =	0	G(100) =	0	G(132) =	0	G(164) =	0	G(196) =	0	G(228) =	0
G(5) =	4	G(37) =	4	G(69) =	5	G(101) =	0	G(133) =	0	G(165) =	0	G(197) =	0	G(229) =	0
G(6) =	88	G(38) =	4	G(70) =	0	G(102) =	0	G(134) =	0	G(166) =	0	G(198) =	0	G(230) =	0
G(7) =	1120	G(39) =	15	G(71) =	1	G(103) =	0	G(135) =	0	G(167) =	0	G(199) =	0	G(231) =	0
G(8) =	1106	G(40) =	0	G(72) =	1	G(104) =	0	G(136) =	0	G(168) =	0	G(200) =	0	G(232) =	0
G(9) =	897	G(41) =	13	G(73) =	0	G(105) =	0	G(137) =	0	G(169) =	0	G(201) =	0	G(233) =	0
G(10) =	1139	G(42) =	5	G(74) =	0	G(106) =	0	G(138) =	0	G(170) =	0	G(202) =	0	G(234) =	0
G(11) =	1616	G(43) =	7	G(75) =	1	G(107) =	0	G(139) =	0	G(171) =	0	G(203) =	0	G(235) =	0
G(12) =	2054	G(44) =	4	G(76) =	0	G(108) =	0	G(140) =	0	G(172) =	0	G(204) =	0	G(236) =	0
G(13) =	2385	G(45) =	9	G(77) =	2	G(109) =	0	G(141) =	0	G(173) =	0	G(205) =	0	G(237) =	0
G(14) =	2415	G(46) =	1	G(78) =	0	G(110) =	0	G(142) =	0	G(174) =	0	G(206) =	0	G(238) =	0
G(15) =	2703	G(47) =	11	G(79) =	0	G(111) =	0	G(143) =	0	G(175) =	0	G(207) =	0	G(239) =	0
G(16) =	1953	G(48) =	0	G(80) =	0	G(112) =	0	G(144) =	0	G(176) =	0	G(208) =	0	G(240) =	0
G(17) =	1649	G(49) =	12	G(81) =	0	G(113) =	0	G(145) =	0	G(177) =	0	G(209) =	0	G(241) =	0
G(18) =	266	G(50) =	4	G(82) =	0	G(114) =	0	G(146) =	0	G(178) =	0	G(210) =	0	G(242) =	0
G(19) =	700	G(51) =	3	G(83) =	0	G(115) =	0	G(147) =	0	G(179) =	0	G(211) =	0	G(243) =	0
G(20) =	645	G(52) =	2	G(84) =	0	G(116) =	0	G(148) =	0	G(180) =	0	G(212) =	0	G(244) =	0
G(21) =	605	G(53) =	2	G(85) =	0	G(117) =	0	G(149) =	0	G(181) =	0	G(213) =	0	G(245) =	0
G(22) =	204	G(54) =	2	G(86) =	0	G(118) =	0	G(150) =	0	G(182) =	0	G(214) =	0	G(246) =	0
G(23) =	279	G(55) =	1	G(87) =	0	G(119) =	0	G(151) =	0	G(183) =	0	G(215) =	0	G(247) =	0
G(24) =	81	G(56) =	6	G(88) =	0	G(120) =	0	G(152) =	0	G(184) =	0	G(216) =	0	G(248) =	0
G(25) =	122	G(57) =	0	G(89) =	0	G(121) =	0	G(153) =	0	G(185) =	0	G(217) =	0	G(249) =	0
G(26) =	111	G(58) =	1	G(90) =	0	G(122) =	0	G(154) =	0	G(186) =	0	G(218) =	0	G(250) =	0
G(27) =	91	G(59) =	2	G(91) =	0	G(123) =	0	G(155) =	0	G(187) =	0	G(219) =	0	G(251) =	0
G(28) =	81	G(60) =	1	(92) =	0	G(124) =	0	G(156) =	0	G(188) =	0	G(220) =	0	G(252) =	0
G(29) =	34	G(61) =	5	G(93) =	0	G(125) =	0	G(157) =	0	G(189) =	0	G(221) =	0	G(253) =	0
G(30) =	50	G(62) =	0	G(94) =	0	G(126) =	0	G(158) =	0	G(190) =	0	G(222) =	0	G(254) =	0
G(31) =	5	G(63) =	4	G(95) =	0	G(127) =	0	G(159) =	0	G(191) =	0	G(223) =	0	G(255) =	0

TABLE 4.3
Display of Histogram Listed in Table 4.2 with Each * Representing 50 Counts

Gray level	
0	I
1	I
2	I
3	I
4	I
5	I
6	I**
7	I**********************
8	I**********************
9	I******************
10	I***********************
11	I********************************
12	I***
13	I**
14	I**
15	I**
16	I***************************************
17	I*********************************
18	I*****
19	I**************
20	I*************
21	I************
22	I****
23	I******
24	I**
25	I**
26	I**
27	I**
28	I**
29	I*
30	I*
31	I
32	I*
33	I*
34	I
35	I
36	I
37	I
38	I
39	I
40	I

1. The leading zero tells us the pixel can only be one of the following: 4, 5, 7, 10, 11, 17, 18
2. Of these the second bit, a one, tells us that the pixel can only be one of the following: 7, 11, 17, 18
3. The third bit, a zero, limits us to: 7, 17, and 18
4. The fourth bit, a zero, tells us that the value is 7 and that the code for the first pixel is completed

TABLE 4.4
Compressed Histogram[a]

Gray level	Bin count	Gray level	Bin count
1	1	33	24
2	4	34	4
3	88	35	4
4	1120	36	15
5	1106	37	13
6	897	38	5
7	1139	39	7
8	1616	40	4
9	2054	41	9
10	2385	42	1
11	2415	43	11
12	2703	44	12
13	1953	45	4
14	1649	46	3
15	266	47	2
16	700	48	2
17	645	49	2
18	605	50	1
19	204	51	6
20	279	52	1
21	81	53	2
22	122	54	1
23	111	55	5
24	91	56	4
25	81	57	3
26	34	58	1
27	50	59	5
28	5	60	1
29	26	61	1
30	36	62	1
31	10	63	2
32	18		

[a]This table is Table 4.2 with the zero-count bins deleted.

5. The fifth bit, a zero, says that the second pixel can only be one of the following: 4, 5, 7, 10, 11, 17, 18.

Continuing this elimination process indicates that 7, 17, 13, 5 are the values of the first four pixels. In terms of the gray levels of Table 4.2, the first four pixels are valued 10, 20, 16, and 8.

The literature abounds with papers on digital image data compression methods. A particularly relevant reference is *Error Free Coding of Multispectral Images* (Wintz & Duan, 1974). A coding for eight frames of LANDSAT

TABLE 4.5
Frequency-Sorted Histogram

Rank	Count	Table 4.4 bin	Rank	Count	Table 4.4 bin
1	2703	12	33	11	43
2	2415	11	34	10	31
3	2385	10	35	9	41
4	2054	9	36	7	39
5	1953	13	37	6	51
6	1649	14	38	5	28
7	1616	8	39	5	38
8	1139	7	40	5	55
9	1120	4	41	5	59
10	1106	5	42	4	2
11	897	6	43	4	34
12	700	16	44	4	35
13	645	17	45	4	40
14	605	18	46	4	45
15	279	20	47	4	56
16	266	15	48	3	46
17	204	19	49	3	57
18	122	22	50	2	47
19	111	23	51	2	48
20	91	24	52	2	49
21	88	3	53	2	53
22	81	21	54	2	63
23	81	25	55	1	1
24	50	27	56	1	42
25	36	30	57	1	50
26	34	26	58	1	52
27	26	29	59	1	54
28	24	33	60	1	58
29	18	32	61	1	60
30	15	36	62	1	61
31	13	37	63	1	62
32	12	44			

achieves an average bit rate of 2.5 bits per pixel per channel is reported for data with entropy of 2.4 bits. (The example showing a Huffman code of 4.09 bits is for data having an entropy of 4.06 bits). Other works on compression include Habibi (1977), Kang, Lee, Chang, and Chang (1977), and O'Neal and Natarajan (1977).

TABLE 4.6
Huffman Code as a Function of Compressed Gray-Level Value as in Table 4.4

1	10100111111000	33	1010011100
2	101000000010	34	101000000001
3	10100100	35	101000000000
4	0001	36	10100111110
5	0000	37	10100111100
6	11010	38	101001100001
7	0100	39	101001111011
8	1011	40	1010011111111
9	1111	41	10100000011
10	001	42	10100001110011
11	011	43	10100110001
12	100	44	10100110011
13	1110	45	1010011111110
14	1100	46	1010011110101
15	1101110	47	1010000001001
16	10101	48	1010000001000
17	01011	49	1010000000111
18	01010	50	10100001110010
19	1101100	51	101001100101
20	1101111	52	10100001110001
21	10100011	53	1010000000110
22	11011011	54	10100001110000
23	11011010	55	101001100000
24	10100101	56	1010011111101
25	10100010	57	1010011110100
26	101000001	58	10100000010111
27	101001101	59	101000011101
28	101001100100	60	10100000010110
29	1010011101	61	10100000010101
30	101000010	62	10100000010100
31	10100001111	63	10100111111001
32	1010000110		

TABLE 4.7
Bits per Pixel as a Function of Compressed Histogram[a]

Gray level	Bits per pixel	Gray level	Bits per pixel
1	14	33	10
2	12	34	12
3	8	35	12
4	4	36	11
5	4	37	11
6	5	38	12
7	4	39	12
8	4	40	13
9	4	41	11
10	3	42	14
11	3	43	11
12	3	44	11
13	4	45	13
14	4	46	13
15	7	47	13
16	5	48	13
17	5	49	13
18	5	50	14
19	7	51	12
20	7	52	14
21	8	53	13
22	8	54	14
23	8	55	12
24	8	56	13
25	8	57	13
26	9	58	14
27	9	59	12
28	12	60	14
29	10	61	14
30	9	62	14
31	11	63	14
32	10		

[a] Gray-level value as in Table 4.6. Frequency-weighted average number of bits per pixel = 4.09220.

Two-Dimensional Hadamard Transform

The Hadamard Transform, which is closely related to the Fourier transform, has, for some applications, advantages over the Fourier transform (Pratt, 1978; Rosenfeld & Kak, 1976). Consider an $n \times n$ array of picture elements X_{ij} where $i, j = 1, 2, \ldots, n$. A transformation of this X_{ij} array into another $n \times n$ array Y_{kl} can be specified by

$$Y_{kl} = \sum_{i=1}^{n} \sum_{i=1}^{n} a_{klik} X_{ij}, \qquad k,l = 1, 2 \ldots, n.$$

Similarly, the inverse transformation gives

$$X_{ij} = \sum_{k=1}^{n} \sum_{l=1}^{n} a_{ijkl} Y_{kl}, \qquad i_j = 1,2 \ldots, n.$$

The particular transform is defined by choosing the as. For the Fourier transformation

$$a_{klij} = \frac{1}{n} \exp[-2\pi\sqrt{-1}(ki + lj)n]$$

and the Hadamard transformation

$$a_{klij} = \frac{1}{n}(-1)^{b(k,l,i,j)}$$

where

$$b(k,l,i,j) = \sum_{h=0}^{\log_2 n - 1} [b_h(k)b_h(l) + b_h(i)b_h(j)]$$

$b_n(*)$ is the hth bit in the binary representation of (*), and n is a power of 2. Stated differently, the Fourier transform uses sines and cosines, whereas the Hadamard transform uses Walsh functions. Figure 4.30 shows the first 16 of these Hadamard transform waveforms.

One advantage of the Hadamard transform over the Fourier transform is computational simplicity. For an $n \times n$ image, the FFT requires $2n^2 \log_2 n$ multiplications and a like number of additions, whereas the Hadamard transform requires just the additions. For image data compression purposes, large images are generally partitioned into a set of 16×16 subimages (Wintz 1972). Sample input and output arrays are shown in Figures 4.31 and 4.32.

The program does not normalize the output. Fortuitously, the inverse Hadamard transform is produced by the same program to within a scale factor.

Some additional reading about Hadamard transformations includes Deutsch and Wann (1972), Harmuth (1968), Kahveci and Hall (1974), Pratt, Kane, and Andrews (1969), Takahashi and Kishi (1980), and Wintz (1972).

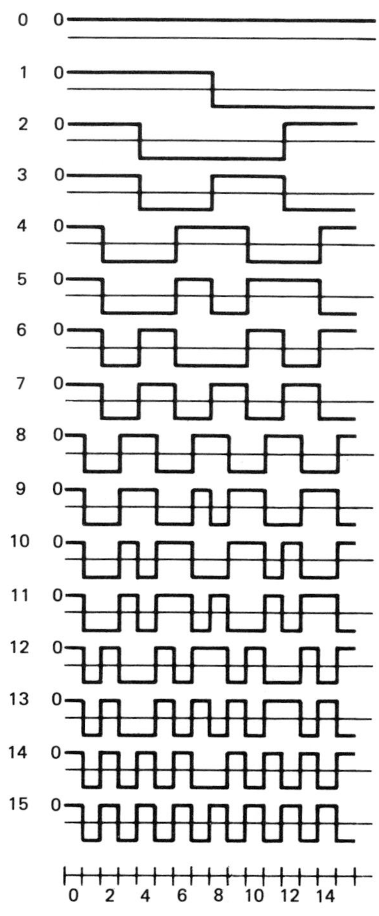

FIGURE 4.30. The first 16 of the Hadamard transform waveforms.

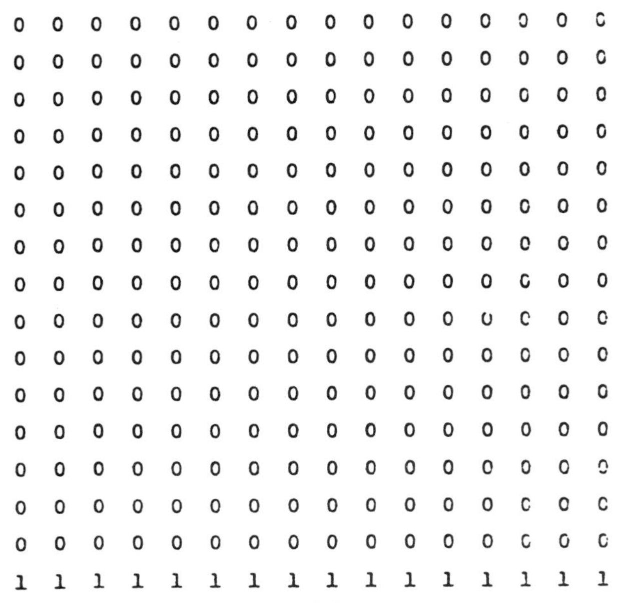

FIGURE 4.31. Sample input array.

Texture Classification

On page 96 multispectral classification was addressed. There each pixel was assigned to a category of land cover based on that pixel's spectral variation. Since the land cover varies in reflectance as a function of wavelength, various categories of land cover can be characterized by a spectral signature for each category, and the pixels are then categorized by matching signatures. A similar classification process can be performed on the basis of the spatial variations in the vicinity of each pixel. These spatial variations are termed *texture,* and an example of texture image classification is given in Chapter 6.

Other examples of texture classification include the Haralick co-occurrence algorithm (Haralick, 1972) and the max–min algorithm developed by Mitchell, Myers, and Boyne (1977). (see also Carton, Weszka, and Rosenfeld, 1974; Hsu, 1979; and Thompson, 1974). It has been shown that classification on the basis of spectral signatures alone is not, in general, as successful as classification on the basis of spectral signatures alone; classification based on both types of information is generally better than using either alone.

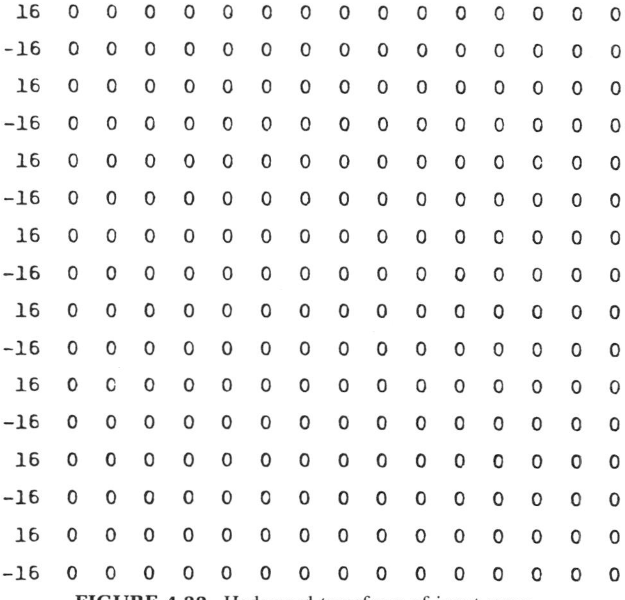

16	0	0	0	0	0	0	0	0	0	0	0	0	0	0	0
-16	0	0	0	0	0	0	0	0	0	0	0	0	0	0	0
16	0	0	0	0	0	0	0	0	0	0	0	0	0	0	0
-16	0	0	0	0	0	0	0	0	0	0	0	0	0	0	0
16	0	0	0	0	0	0	0	0	0	0	0	0	0	0	0
-16	0	0	0	0	0	0	0	0	0	0	0	0	0	0	0
16	0	0	0	0	0	0	0	0	0	0	0	0	0	0	0
-16	0	0	0	0	0	0	0	0	0	0	0	0	0	0	0
16	0	0	0	0	0	0	0	0	0	0	0	0	0	0	0
-16	0	0	0	0	0	0	0	0	0	0	0	0	0	0	0
16	0	0	0	0	0	0	0	0	0	0	0	0	0	0	0
-16	0	0	0	0	0	0	0	0	0	0	0	0	0	0	0
16	0	0	0	0	0	0	0	0	0	0	0	0	0	0	0
-16	0	0	0	0	0	0	0	0	0	0	0	0	0	0	0
16	0	0	0	0	0	0	0	0	0	0	0	0	0	0	0
-16	0	0	0	0	0	0	0	0	0	0	0	0	0	0	0

FIGURE 4.32. Hadamard transform of input array.

Cloud Removal

Earlier digital image mosaics have been discussed (page 72). The concepts covered can be extended for the reduction or removal of cloud cover from digital imagery.

Assume there are two images, A and B, of the same terrain, acquired at different times. Assume, also, that A has been registered to B. If both A and B exhibit substantial cloud cover, then a new composite image, C, can be generated from A and B. C has reduced cloud cover, provided that the clouds are distributed differently in A and B.

The algorithm is in two steps:

1. $G_C = G_A$
2. If $G_C > T$, $G_C = G_B$

Said differently, the new image is made up of the pixels in image A that have a gray level that is less than some specified threshold T; if a pixel from image A exceeds T in gray level, then it is considered a cloud pixel, and the corresponding pixel from image B is inserted instead.

REFERENCES

Amin, A. T. An algorithm for gray-level transformations in digital images. *IEEE Transactions on Computers,* 1977, *26,* 1158–1161.

Andrews, H. C. *Computer techniques in image processing.* New York: Academic Press, 1970.

Andrews, H. C. Digital Fourier transforms as means for scanner evaluation. *Applied Optics,* 1974, *13,* 146–149.

Andrews, H. C. Digital image restoration: a survey. *Computer,* 1974, *7,* 36–45.

Andrews, H. C., & Hunt, B. R. *Digital image restoration.* Englewood Cliffs, New Jersey: Prentice-Hall, 1977.

Arcelli, C., & Dibaja, C. S. Medial lines and figure analysis. Paper presented at the Institute for Electrical and Electronic Engineers 5th international conference on pattern recognition, Miami Beach, Florida, December 1980.

Barnea, D., & Silverman, H. A class of algorithms for fast digital image registration. *IEEE Transactions on Computers,* 1972, *C-21,* 179–186.

Bergland, G. D. Fast Fourier transform hardware implementations—an overview. *IEEE Transactions on Audio and Electroacoustics,* 1969, *AU-17,* 104–108.

Bergland, G. D. Fast Fourier transform hardware implementations—a survey. *IEEE Transactions on Audio and Electroacoustics,* 1969, *AU-17,* 109–119.

Bergland, G. D. A guided tour of the fast Fourier transform. *IEEE Spectrum,* 1969, *6,* 41–52.

Billingsley, F. C. Applications of digital image processing. *Applied Optics,* 1970, *9,* 289–299.

Boulter, J. F. Use of two dimensional Fourier transforms for image processing and analysis. NASA Document N76-11765, National Aeronautics and Space Administration, Washington, D.C., 1976.

Boulter, J. F. Interactive digital image restoration and enhancement. *Computer Graphics and Image Processing,* 1979, *9,* 301–312.

Buijs, H. L., Pomerleau, A., & Fournier, M. Implementation of a fast Fourier transform (FFT) for image processing applications. *IEEE Transactions on Acoustics, Speech and Signal Processing,* 1974, *ASSP-22,* 420–424.

Carton, E. J., Weszka, J. S., & Rosenfeld, A. Some basic texture analysis techniques. University of Maryland Computer Science Center (Tech. Rep. TR-288), 1976.

Champeney, D. C. Fourier transforms and their physical applications. New York: Academic Press, 1973.

Cochran, W. T., Cooley, J. W., Favin, D. L., Helms, H. D., Kaenel, R. A., Lang, W. W., Maling, G. C., Nelson, D. E., Rader, C. M., & Welch, P. D. What is the fast Fourier transform?. *IEEE Transactions on Audio and Electroacoustics,* 1967, *AU-15,* 45–55.

Cooley, J. W., Lewis, P. A., & Welch, P. D. Applications of the fast Fourier transform to computation of Fourier integrals, Fourier series and convolution integrals. *IEEE Transactions on Audio and Electroacoustics,* 1967, *AU-15,* 79–84.

Cooley, J. W., Lewis, P. A., & Welch, P. D. The finite Fourier transform. *IEEE Transactions on Audio and Electroacoustics,* 1969, *AU-17,* 77–85.

Corinthios, M. J., Smith, K. C., & Yen, J. L. A parallel radix-4 fast Fourier transform computer. *IEEE Transactions on Computers,* 1975, *C-24,* 80–92.

Davis, L. S. A survey of edge detection techniques. *Computer Graphics and Image Processing,* 1975, *4,* 248–270.

Deutsch, E. S., & Wann, Z. F. On the use of the Hadamard transform in digital image matching. University of Maryland Computer Science Center (Tech. Rep. TR-216), 1972.

Duda, R. O., & Hart, P. E. Use of the Hough transformation to detect lines and curves in pictures. *Communications of the Association for Computing Machinery,* 1972, *15,* 11.

Frei, W. Image enhancement by histogram hyperbolization. *Computer Graphics and Image Processing,* 1977, *6,* 286–294.

Frieden, B. R. Restoring with maximum likelihood and maximum entropy. *Journal of the Optical Society of America,* 1972, *62,* 511–518.

Gallagher, N. C., & Liu, B. Optimum Fourier transform division filters with magnitude constraint. *Journal of the Optical Society of America,* 1974, *64,* 1227–1236.

Gallager, R. G. Variations on a theme by Huffman. *IEEE Transactions on Information Theory,* 1978, *24,* 668–674.

Gottlieb, P., & Lorenzo, J. Parallel data streams and serial arithmetic for fast Fourier transform processors. *IEEE Transactions on Acoustics, Speech and Signal Processing,* 1974, *ASSP-22,* 111–117.

Habibi, A. Special issue on image bandwidth compression. *IEEE Transactions on Communications,* 1977, *25,* 1249–1440.

Hale, J. A. G., & Saraga, P. Review: Digital image processing. *Opto Electronics,* 1974, *6,* 338–348.

Hall, D. J., Endlich, R. M., Wolf, D. E., & Brain, A. E. Objective methods for registering landmarks and determining cloud motions from satellite data. *IEEE Transactions on Computers,* 1972, *C-21,* 768.

Hall, E. L. A comparison of computations for spatial frequency filtering. *Proceedings of the IEEE, 60,* 887–891.

Haralick, R. Texture tone study—Second interim technical report. University of Kansas Center for Research, 1972.

Harmuth, H. F. A generalized concept of frequency and some applications. *IEEE Transactions on Information Theory,* 1968, *II-14,* 375.

Helms, H. D. Fast Fourier transform method of computing difference equations and simulating filters. *IEEE Transactions on Audio and Electroacoustics,* 1967, *AU-15,* 85–90.

Helstrom, C. W. Image restoration by the method of least squares. *Journal of the Optical Society of America,* 1967, *57,* 297–303.

Herbst, N. M., & Will, P. M. An experimental laboratory for pattern recognition and signal processing. *Communications of the Association for Computing Machinery,* 1972, *15,* 231–244.

Horn, B. K. P., & Woodham, R. J. Destriping satellite images. Proceedings: Image Understanding Workshop, 1978, Cambridge, Mass., pp. 56–63.

Hough, P. V. C. Method and means for recognizing complex platterns. U.S. Patent 3069654, 1962.

Hsu, S. Y., The Mahalanobis classifier with the generalized inverse approach for automated analysis of imagery texture data. *Computer Graphics and Image Processing,* 1979, *9,* 117–134.

Hunt, R. B. The application of constrained least squares estimation to image restoration by digital computer. *IEEE Transactions on Computers,* 1973, *C-22,* 805–812.

Huang, T. S., Schreiber, W. F., & Tretiak, O. J. Image processing. *Proceedings of the IEEE,* 1971, *59,* 1586–1609.

Hummel, R. Image enhancement by histogram transformation. *Computer Graphics and Image Processing,* 1977, *6,* 184–195.

Hung, S. H. Y. A generalization of DPCM for digital image compression. *IEEE Transactions on Pattern Analysis and Machine Intelligence,* 1979, *2,* 100–109.

Jarvis, R. A. An interactive minicomputer laboratory for graphics, image processing and pattern recognition. *Computer,* 1974, *6,* 49–60.

Kahveci, A. E., & Hall, E. L. Sequency domain design of frequency filters. *IEEE Transactions on Computers,* 1974, *C-23,* 976.

Kang, A. N. C., Lee, R. C. T., Chang, C. L., & Chang, S. K. Storage reduction through minimal spanning trees and spanning forests. *IEEE Transactions on Computers,* 1977, *26,* 425–434.

Keshaven, H. R., & Srinath, M. D. Interpolative models in restoration and enhancement of noisy images. *IEEE Transactions on Acoustics, Speech and Signal Processing,* 1977, *25,* 525–534.

Lahart, M. J. Maximum-likelihood restoration of nonstationary imagery. *Journal of the Optical Society of America,* 1974, *64,* 17–22.

Lam, C. F., & Hoyt, R. R. High speed image correlation for change detection. *NAECON Proceedings,* 1972, *24,* 30–37.

Landgrebe, D. The development of a spectral/spatial classifier for earth observational data. Proceedings of the IEEE Pattern Recognition and Image Processing Conference, 1978, pp. 470–475.

Landgrebe, D. Monitoring the earth's resources from space- can you really identify crops by satellite? Proceedings of the National Computer Conference, 1979, *48,* 233.

Lev, A., Zucker, S. W., & Rosenfeld, A. Iterative enhancement of noisy images. *IEEE Transactions on Systems, Man and Cybernetics,* 1977, *7,* 435–442.

Lindenlaub, J. Remote sensing analysis: a basic preparation (LARS Information Note 110471), Purdue University Laboratory for Applications of Remote Sensing, 1972.

Lipkin, B. S., & Rosenfeld. Picture processing and psychopictorics. New York: Academic Press, 1970.

McGlamery, B. L. Restoration of turbulence-degraded images. *Journal of the Optical Society of America,* 1967, *57,* 293–297.

Milgram, D. L. Computer methods for creating photomosaics. *IEEE Transactions on Computers,* 1975, *C-24,* 1113.

Milgram, D. L. Adaptive techniques for photomosaiking. *IEEE Transactions on Computers,* 1977, *C-26,* 1175.

Mitchell, O. R., Myers, C. R., & Boyne, W. A max–min measure for image texture analysis. *IEEE Transactions on Computers,* 1977, *C-26,* 408–414.

Nagel, R. N., & Rosenfeld, A. Ordered search techniques in template matching. *Proceedings of the IEEE,* 1972, *60.*

Nagy, G. Digital image processing activities in remote sensing for earth resources. *Proceedings of the IEEE,* 1972, *60,* 1177–1200.

Nahi, N. E. Role of recursive estimation in statistical image enhancement. *Proceedings of the IEEE,* 1972, *60,* 872–877.

Nevatia, R. A color edge detector and its use in scene segmentation. *IEEE Transactions on Systems, Man and Cybernetics,* 1977, *7,* 820–826.

O'Handley, D. A., & Green, W. B. Recent developments in digital image processing at the image processing laboratory at the Jet Propulsion Laboratory. *Proceedings of the IEEE,* 1972, *60,* 821–828.

O'Neal, J. B., & Natarajan, T. R. Coding isotropic images. *IEEE Transactions on Information Theory,* 1977, *23,* 697–707.

Panda, D. P. Nonlinear smoothing of pictures. *Computer Graphics and Image Processing,* 1978, *8,* 259.

Panda, D. P., & Kak, A. C. Recursive least squares smoothing of noise in images. *IEEE Transactions on Acoustics, Speech and Signal Processing,* 1977, *25,* 520–524.

Pratt, W. K. Digital image processing. New York: Wiley, 1978.

Pratt, W. K., Kane, J., & Andrews, H. C. Hadamard transform image coding. *Proceedings of the IEEE,* 1969, *13,* 58.

Pratt, W. K. (Ed.). Image transmission techniques. New York: Academic Press, 1979.

Ramapriyan, H. K. Data handling for the geometric correction of large images. *IEEE Transactions on Computers,* 1977, *C-26,* 1163–1167.

Richardson, W. H. Baysian-based iterative method for image restoration. *Journal of the Optical Society of America,* 1972, *62,* 55–59.

Rosenfeld, A. Picture processing by computer. New York: Academic Press, 1969.

Rosenfeld, A., & Kak, A. C. Digital picture processing. New York: Academic Press, 1976.

Rosenfeld, A. Picture processing. *Computer Graphics and Image Processing,* 1972, *1,* 394–416.

Schacter, B. J., Lev, A., Zucker, S. W., & Rosenfeld, A. An application of relaxation methods to edge reinforcement. *IEEE Transactions on Systems, Man and Cybernetics,* 1977, *7,* 813–816.

Shapiro, L. G., & Haralick, R. M. Algorithms for inexact matching. Paper presented at the IEEE 5th international conference on pattern recognition, Miami Beach, Florida, December, 1980.

Shaw, G. B. Local and regional edge detectors: some comparisons. *Computer Graphics and Image Processing,* 1979, *9,* 135–149.

Sondhi, M. M. Image restoration: The removal of spatially invariant degradations. *Proceedings of the IEEE,* 1972, *60,* 842–853.

Stefanelli, R. S., & Rosenfeld, A. Some parallel thinning algorithms for digital pictures. *Journal of the Association of Computing Machinery,* 1971, *18,* 255–264.

Swain, P. Pattern recognition: a basis for remote sensing data analysis. (LARS Information Note 11572) Purdue University Laboratory for Applications of Remote Sensing, 1972.

Takahashi, K., & Kishi, T. Feature extraction method by specialized Hadamard transform. Paper presented at the IEEE 5th international conference on pattern recognition, Miami Beach, Florida, December 1980.

Thompson, W. B. The role of texture in computerized scene analysis, University of Southern California, (USC IPI Report 550) 1974.

Webber, W. F. Techniques for image registration, LARS. Proceedings on Machine Processing of Remotely Sensed Data, 1973, 1b1–ib7.

Wechsler, H., & Kidode, M. A new edge detection technique and its implementation. *IEEE Transactions on Systems, Man and Cybernetics,* 1977, *7,* 827–836.

Wintz, P. A. Transform picture coding. *Proceedings of the IEEE,* 1972, *60,* 809.

Wintz, P. A., & Duan, J. R. Summary report of research, Error free coding of multispectral images, (TR-EE 74-22), Purdue University, 1974.

5 Application Examples

This chapter discusses examples of how digital image processing algorithms like those discussed previously have been applied to real imagery for specific objectives.

KENTUCKY WATER IMPOUNDMENT SURVEY

After Congress passed the Dam Safety Act (1973), the state governments were obliged to assure the federal government that all water impoundments were not sources of flooding risk. Permanent dams did not present an inspection challenge, but temporary dams, for example those formed by mine slag heaps, were generally unknown in number and location. The Commonwealth of Kentucky arranged for a survey to be performed using Landsat imagery and digital analysis. Band 7, in which water pixels are very dark, proved to have the greatest utility. The process is described graphically in figures 5.1*a* and 5.1*b*.

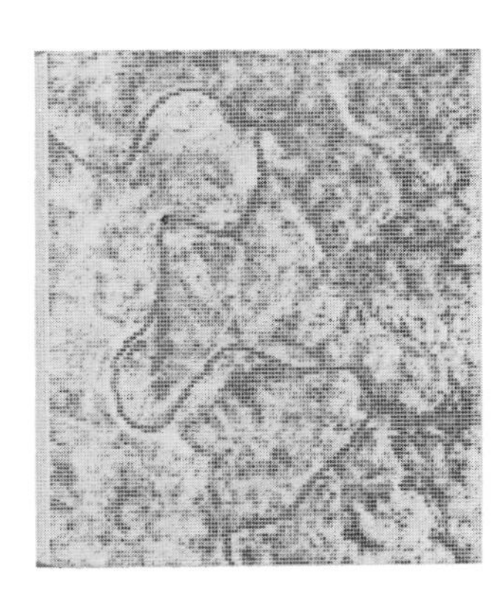

FIGURE 5.1a. Kentucky water impoundment experiment. The ability to effectively measure the area extent and location of surface water bodies from remotely sensed imagery is of importance in many areas of the world. Such evaluations provide necessary information for analysis of regional water resources, agricultural practices, and land use policy. The application described below was performed for the commonwealth of Kentucky and required the reporting of all detected surface water bodies by size and location, for use in dam safety studies. The accurate positioning of water bodies by geographic coordinators is accomplished by an affine transformation which maps column and row indices from the imagery into latitude and longitude. Because the coefficients of this transformation are unique to each image, their determination is accomplished through the use of ground control points (GCPs). Working with a digitally produced erts image and a usgs 7½ minute quad sheet, a shade print is produced at the same scale as the quad sheet. The photointerpreter selects and circles corresponding features (GCP) in the quad sheet and shade print for coordinate determination. This process is repeated until the affine transform can be accurately defined.

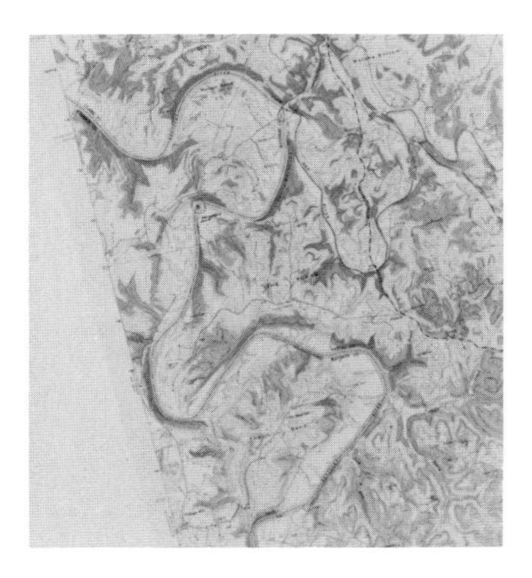

FIGURE 5.1b. Variations in the appearance of water within the erts image require the determination of the spectral response of water. Because of the variability of the spectral response of water based on the environment in which it occurs (eg, depth, amount of sediment, or width of a stream) an iterative analysis of the data is performed in order to define the spectral signal of water which yields minimal error. This procedure is enhanced by the use of specially constructed shade prints displaying those pixels found to contain water. At the conclusion of the process, a tabular display of the water inventory is produced.

DIGITAL LANDSAT PROCESSING TO ASSESS NEW YORK BIGHT ACID DUMP[1]

As large metropolitan centers have developed, the quantities of waste materials deposited in coastal oceans have increased. They are making a recognizable impact on the marine environment. What impact such environmental changes may have remains to be determined.

Besides having a large amount of marine traffic with its attendant pollution problems, the New York Bight area has been used for deliberate coastal marine waste disposal for well over a century (Johnson, 1980). Waste disposal includes sewage, sludge, toxic chemicals, and ocean outfalls. The environmental impact of these wastes and their movement, dispersion, and accumulation have not been studied in detail, nor were systematic investigations seriously considered before 1968. The National Oceanic and Atmospheric Administration (NOAA) initiated an effort in 1972 to develop a better understanding of the New York Bight ecosystem. This program, called the Marine Ecosystems Analysis (MESA) Program, has been studying physical, chemical, and biological interactions within the New York Bight.

Numerous federal agencies, state governments, and local municipalities are involved in the management of the New York Bight. These agencies require information about environmentally related characteristics and phenomena that

[1]This section originally appeared as Hord, R. M., & Macomber, R., Digital Landsat processing to assess New York Bight acid dump, *Proceedings of the American Society of Photogrammetry*, 1979, *43*, 638.

can contribute realistically to routine decision making and to practical coastal management. To show the benefits of remote sensing data acquisition applied to studies of the coastal marine ecosystem in the New York Bight area and of remote-sensing-based information products that can be provided to area users, a demonstration using remote sensor records for the study of circulation, dispersion, and related environmental characteristics was undertaken. This effort was sponsored by the Spacecraft Oceanography Group of the NOAA (Macomber, Mairs, & Stanczuk, 1974).

Operational Phase

The operational phase of this remote sensing study required about 4 months of planning and preparation, 3 weeks of field equipment preparation and setup, and 1 day of concentrated data collection. A basemap of the New York Bight marine environment was prepared from Landsat imagery enlarged to 1:500,000 scale. Locations of approved dump sites for acid waste, sewer sludge, cellar dirt, and mud were indicated in relationship to Ambrose Light on this product (see Figure 5.2).

Seven small boat transects with from six to seven stations each were laid out on a nautical chart. Two helicopters were used for dye implants on the *outer* New York Bight. Seven boats were procured for dye drop and water sampling transects in the *inner* New York Bight. By the time of the Landsat overpass at 10:35 A.M. EST, 7 April 1973, all dye implants had been streaming for at least 2 hours. A water sample was taken at each station and the time recorded in the boat log. These samples were to be analyzed by the Coulter Counter, an instrument that measures particle size and distribution in a water sample. Other measurements were made as well.

Landsat Data Computer Enhancement and Analysis

PRELIMINARY DATA ENHANCEMENT

One of the objectives of the New York Bight study was to determine the dye implant specifications necessary to create a streamer large enough and bright enough to be imaged by Landsat. The boat-deployed dye tubes used in this experiment were 3 in. (7.6 cm) in diameter and 3 ft long (.9 m) and were filled (packed) with fluoroscein dye in powder form. The helicopter-deployed dye tubes were 4 in. × 5 ft (10 cm × 1.5 m) and contained a 60–40 fluoroscein dye-wax pellet.

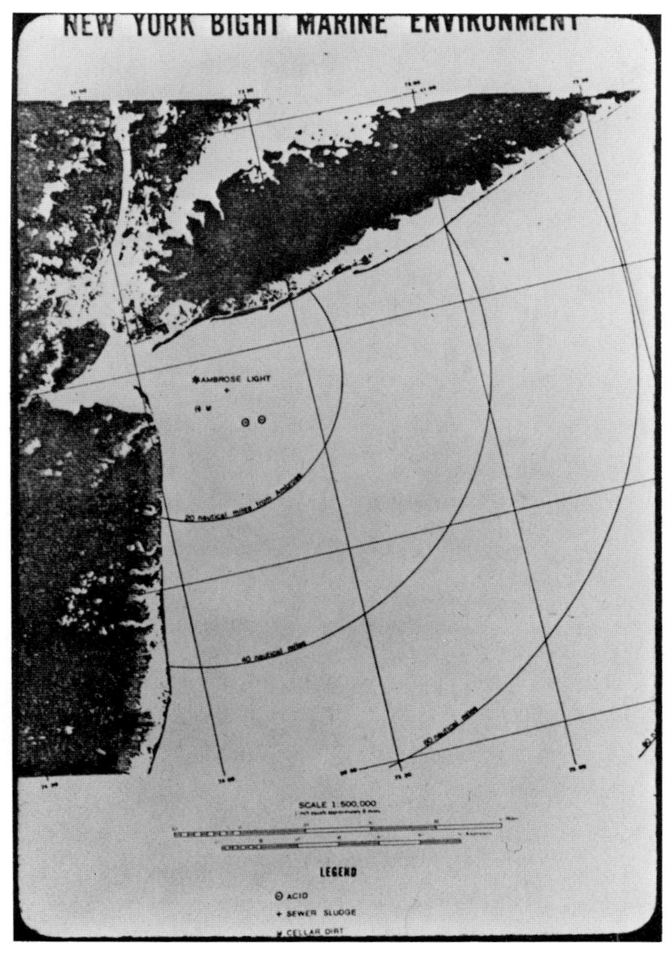

FIGURE 5.2. Locations of approved dump sites for acid waste, sewer sludge, cellar dirt, and mud are indicated in relationship to Ambrose Light. A dot inside a circle = acid; a plus mark = sewer sludge; a "W" = cellar dirt; a dot inside parentheses = mud and one man stone.

On examination of the routinely produced NASA Landsat 9 × 9 inch prints, it was not possible to recognize or locate either of these two types of dye streamers. It was thus initially concluded that the area of the dye and the contrast difference between the bright green dye streamer and the blue-green background water was too small to be imaged by Landsat. However, computer-enhanced prints generated from the NASA-supplied CCTs did image successfully some 23 dye streamers of both types in the Hudson River waters and the clearer coastal ocean waters. Interpretation of the enhanced Landsat images was verified by comparison with 1:40,000 color airphotographs.

Processing to produce these enhanced Landsat images included a four-band Reformat Program to satisfy data handling compatibility requirements; a Window Program that enhanced contrast by redistributing the original histogram of gray levels to a more even distribution within the full dynamic range of intensity levels (0–255); and a Squaring Program designed to correct the inherent geometric infidelity of the data as collected by the satellite. A more exacting registration program, using known groundpoint positional information, was not deemed necessary at this stage of the processing.

A Litton printer (film recorder) compatible tape was then generated from the processed data of band 4, and a black and white enhanced photographic print was produced. This processing provided the photointerpreter with an improved contrast image of band 4. Thus, subtle intensity differences of one or two intensity levels from pixel to pixel were recognizable. A similar sequence of processing was performed on bands 5, 6, and 7. Analysis of band 5 (red band) yielded different information from that apparent in band 4 (green band) due to greater depth penetration and greater haze susceptibility in the

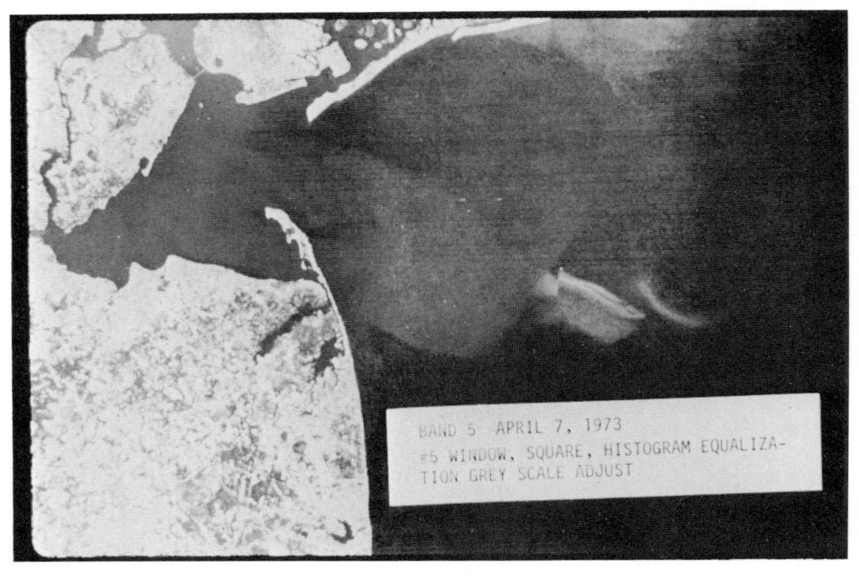

FIGURE 5.3. Band 5, enhanced image. The green dye streamers were not imaged by the red-sensitive band 5.

green-sensitive band. As expected, the green dye streamers were not imaged by the red-sensitive band 5 (see Figure 5.3). Bands 6 and 7 (infrared) showed little information over water and were not considered further.

BAND 4 CLASSIFICATION OF SCENE COMPONENTS

To discriminate the various water masses, dye streamers, and the acid dump dispersion, intensity slice techniques were employed. This was a supervised, photointerpreter-directed, computer-executed discrimination. A computer-generated shade print that character encodes the intensity level of each pixel was made. The interpreter could then identify the character codes of any scene component. The intensity levels associated with each one of the classes were determined by careful inspection of the character code set of the computer-generated shade print. Every line and every pixel was printed, yielding a 5 × 8 ft (1.5 × 2.4 m) shade print. Shorelines, islands, and control points such as Ambrose Light were outlined on the shade print with the aid of U.S. Coast and Geodetic Nautical Charts. Areas interpreted from the band 4 enhanced Litton print as falling into one of the scene component classifications were located on the shade print and their intensity level ranges were noted from the character set to yield Table 5.1.

Using this information, an intensity slice Litton print was made for each of the seven separate scene components. On each print, the intensity level range of the scene component of interest was assigned to density level 255 and thus imaged white. All other *water* area intensity levels in the original data were

TABLE 5.1
Landsat Band 4 Scene Components and Associated Intensity Levels

Scene component classification	Intensity level
1. Dye implant streamers in Hudson River Discharge Plume	29–31
2. Fresh acid dump (10 hr old)	29–33
3. Dispersed acid dump (24 hr old)	23–28
4. Hudson River Discharge Plume (tide between maximum ebb and slack water)	26–30
5. Turbid sediment-laden coastal waters	23–27
6. Clearer coastal–ocean upwelled waters	20–22
7. Dye implant streamers in coastal waters	26–28

The intensity levels associated with each of these classes were determined from careful inspection of the character-code set of a computer-generated shade print of the Landsat 7 April 1973 overpass of the New York Bight.

assigned to level zero and thus imaged black. Land area intensity levels were assigned an intermediate shade of gray, thus showing the land outline as a reference (see Figures 5.4–5.8).

CALCULATION OF INTENSITY LEVEL OF DYE STREAMERS IN CLEARER COASTAL WATERS

The only dye streamers readily identifiable on the computer-generated shade print were the two on boat six transect that lie halfway between Crookes Point and Rockaway Point in the middle of the Hudson River Discharge Plume. It was not possible to locate dye streamers in the clearer coastal waters directly in the shade print. After the two readily visible and positively identified dye markers were located on the shade print, the intensity levels of the individual pixels that made up the dye streamer image were determined to be intensity levels 29–31. This information was used to specify an intensity slice on the Litton recorder that rendered those intensity levels as black, all other levels in the water as white, and the land mass as an intermediate shade of gray. This intensity slice was successful in discriminating these two dye

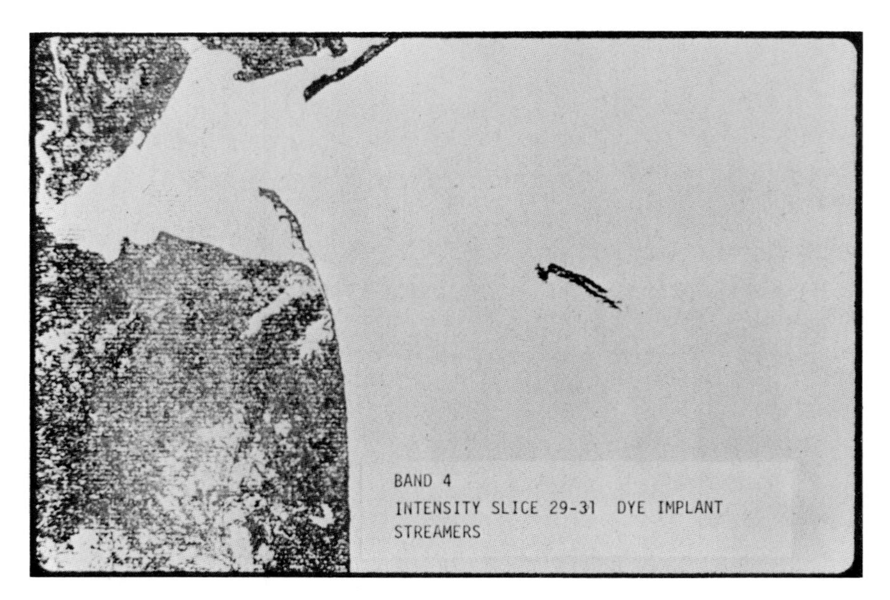

FIGURE 5.4. Band 4, intensity slice 1: Dye implant streamers in Hudson River Discharge Plume. Intensity slice of levels 29–31 (black rendering).

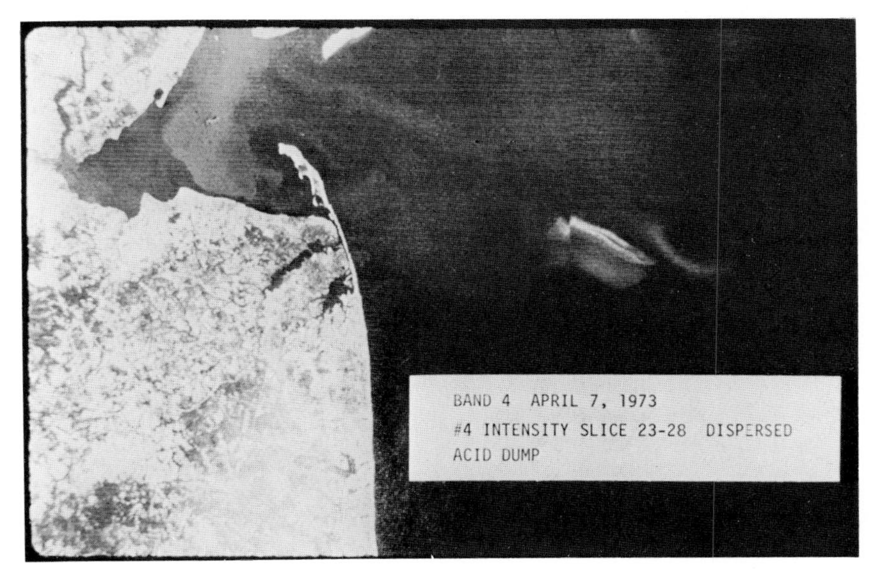

BAND 4 APRIL 7, 1973
#4 INTENSITY SLICE 23-28 DISPERSED
ACID DUMP

FIGURE 5.5. Band 4, intensity slice 2: Fresh acid dump (10 hr old). Intensity slice of levels 29–33 (white rendering).

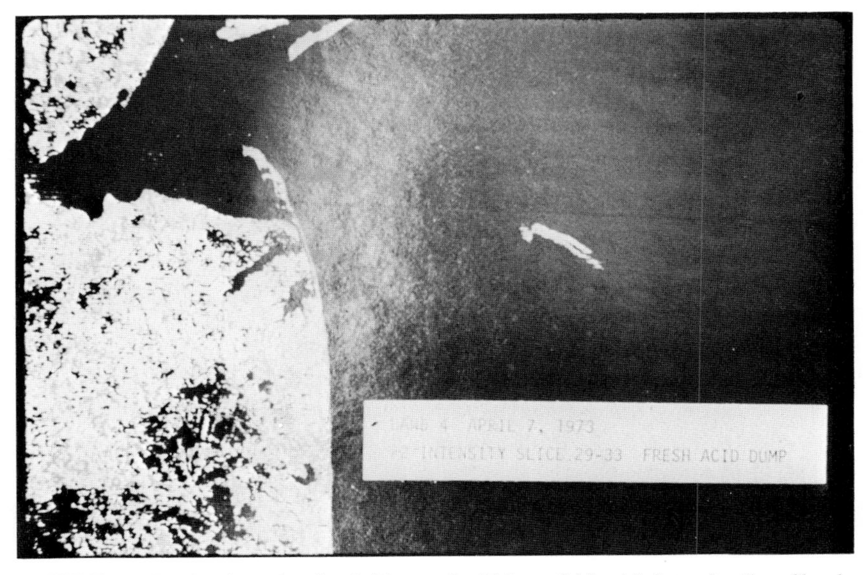

FIGURE 5.6. Band 4, intensity slice 3: Dispersed acid dump (24 hr old). Intensity slice of levels 23–28 (white rendering).

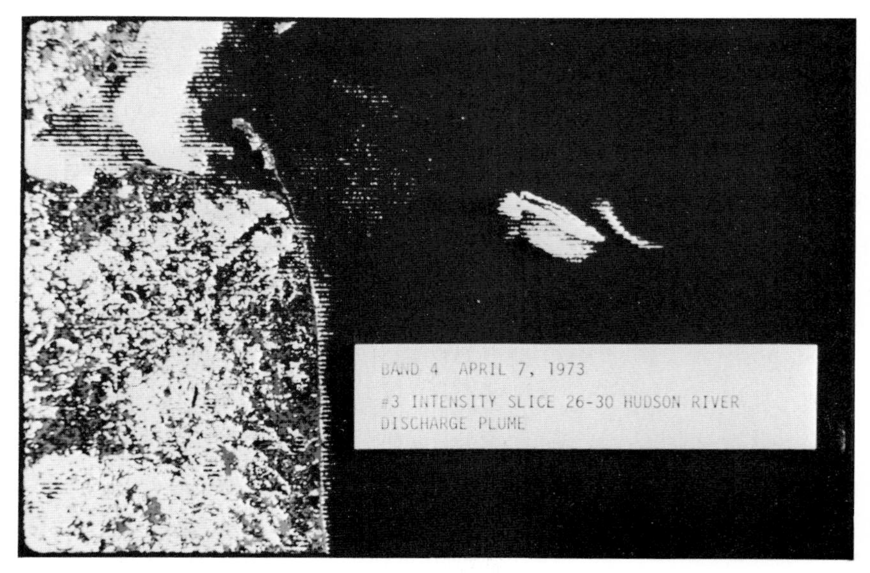

FIGURE 5.7. Band 4, intensity slice 4: Hudson River Discharge Plume. Intensity slice of levels 26–30 (white rendering).

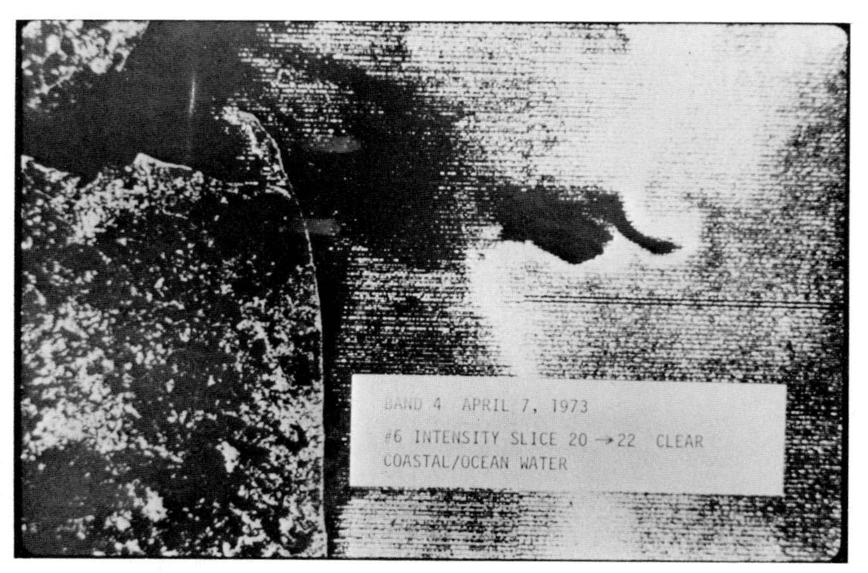

FIGURE 5.8. Band 4, intensity slice 6: Clearer coastal–ocean upwelled waters. Intensity slice of levels 20–22 (white rendering).

G(0) =	0	G(32) =	867	G(64) =	2	G(96) =	0	G(128) =	0	G(160) =	0	G(192) =	0
G(1) =	0	G(33) =	1640	G(65) =	0	G(97) =	0	G(129) =	0	G(161) =	0	G(193) =	0
G(2) =	0	G(34) =	1561	G(66) =	1	G(98) =	0	G(130) =	0	G(162) =	0	G(194) =	0
G(3) =	0	G(35) =	640	G(67) =	1	G(99) =	0	G(131) =	0	G(163) =	0	G(195) =	0
G(4) =	0	G(36) =	454	G(68) =	2	G(100) =	0	G(132) =	0	G(164) =	0	G(196) =	0
G(5) =	0	G(37) =	721	G(69) =	7	G(101) =	0	G(133) =	0	G(165) =	0	G(197) =	0
G(6) =	0	G(38) =	464	G(70) =	1	G(102) =	0	G(134) =	0	G(166) =	0	G(198) =	0
G(7) =	0	G(39) =	394	G(71) =	0	G(103) =	0	G(135) =	0	G(167) =	0	G(199) =	0
G(8) =	0	G(40) =	143	G(72) =	3	G(104) =	0	G(136) =	0	G(168) =	0	G(200) =	0
G(9) =	0	G(41) =	173	G(73) =	0	G(105) =	0	G(137) =	0	G(169) =	0	G(201) =	0
G(10) =	0	G(42) =	145	G(74) =	1	G(106) =	0	G(138) =	0	G(170) =	0	G(202) =	0
G(11) =	0	G(43) =	76	G(75) =	2	G(107) =	0	G(139) =	0	G(171) =	0	G(203) =	0
G(12) =	0	G(44) =	173	G(76) =	0	G(108) =	0	G(140) =	0	G(172) =	0	G(204) =	0
G(13) =	0	G(45) =	90	G(77) =	1	G(109) =	0	G(141) =	0	G(173) =	0	G(205) =	0
G(14) =	0	G(46) =	61	G(78) =	0	G(110) =	0	G(142) =	0	G(174) =	0	G(206) =	0
G(15) =	0	G(47) =	30	G(79) =	0	G(111) =	0	G(143) =	0	G(175) =	0	G(207) =	0
G(16) =	0	G(48) =	94	G(80) =	0	G(112) =	0	G(144) =	0	G(176) =	0	G(208) =	0
G(17) =	2	G(49) =	42	G(81) =	0	G(113) =	0	G(145) =	0	G(177) =	0	G(209) =	0
G(18) =	387	G(50) =	20	G(82) =	1	G(114) =	0	G(146) =	0	G(178) =	0	G(210) =	0
G(19) =	14313	G(51) =	65	G(83) =	0	G(115) =	0	G(147) =	0	G(179) =	0	G(211) =	0
G(20) =	25784	G(52) =	23	G(84) =	0	G(116) =	0	G(148) =	0	G(180) =	0	G(212) =	0
G(21) =	3067	G(53) =	11	G(85) =	0	G(117) =	0	G(149) =	0	G(181) =	0	G(213) =	0
G(22) =	5049	G(54) =	44	G(86) =	0	G(118) =	0	G(150) =	0	G(182) =	0	G(214) =	0
G(23) =	12311	G(55) =	28	G(87) =	0	G(119) =	0	G(151) =	0	G(183) =	0	G(215) =	0
G(24) =	11189	G(56) =	10	G(88) =	0	G(120) =	0	G(152) =	0	G(184) =	0	G(216) =	0
G(25) =	9898	G(57) =	23	G(89) =	0	G(121) =	0	G(153) =	0	G(185) =	0	G(217) =	0
G(26) =	16867	G(58) =	2	G(90) =	0	G(122) =	0	G(154) =	0	G(186) =	0	G(218) =	0
G(27) =	11267	G(59) =	11	G(91) =	0	G(123) =	0	G(155) =	0	G(187) =	0	G(219) =	0
G(28) =	2666	G(60) =	9	G(92) =	0	G(124) =	0	G(156) =	0	G(188) =	0	G(220) =	0
G(29) =	1083	G(61) =	2	G(93) =	0	G(125) =	0	G(157) =	0	G(189) =	0	G(221) =	0
G(30) =	2151	G(62) =	0	G(94) =	0	G(126) =	0	G(158) =	0	G(190) =	0	G(222) =	0
G(31) =	923	G(63) =	4	G(95) =	0	G(127) =	1	G(159) =	0	G(191) =	0	G(223) =	0

FIGURE 5.9. Histogram table, band 4, 7 April 1973. The vertical columns of numbers in parentheses are the intensity levels (0–255 total, 0–223 shown) of reflected energy received by Landsat. The figure to the right of the equal signs is the tally of the total number of pixels in the scene window that received that intensity level of energy. This window is predominantly a water scene, as indicated by the large number of pixels present in the lower intensity levels (18–27). Levels 29–33 are the acid waste dump. Levels 34 and above are land areas.

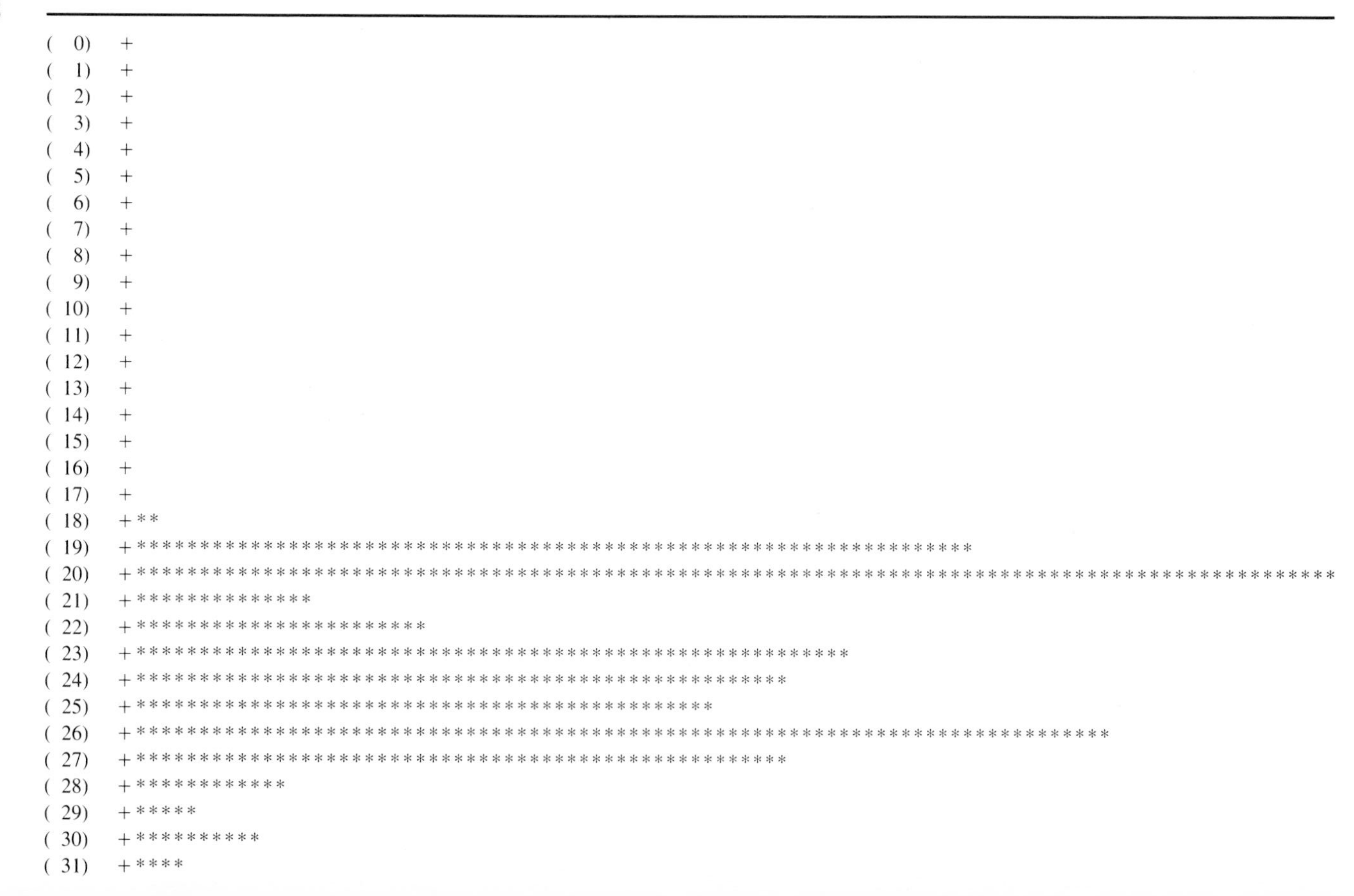
(0)
(1)
(2)
(3)
(4)
(5)
(6)
(7)
(8)
(9)
(10)
(11)
(12)
(13)
(14)
(15)
(16)
(17)
(18)
(19)
(20)
(21)
(22)
(23)
(24)
(25)
(26)
(27)
(28)
(29)
(30)
(31)

```
( 32)  +****
( 33)  +********
( 34)  +*******
( 35)  +***
( 36)  +**
( 37)  +***
( 38)  +**
( 39)  +**
( 40)  +*
( 41)  +*
( 42)  +*
( 43)  +
( 44)  +*
( 45)  +
( 46)  +
( 47)  +
( 48)  +
( 49)  +
( 50)  +
( 51)  +
( 52)  +
( 53)  +
( 54)  +
( 55)  +
( 56)  +
( 57)  +
( 58)  +
( 59)  +
```

FIGURE 5.10. Intensity level versus number of pixels that have a given intensity level. The distribution of intensity levels in the scene window is presented in histogram form. The histogram equalization gray-scale adjust used to enhance the contrast of the output essentially flattens this histogram and spreads it evenly over the 0–255 intensity level range. Note that in the original histogram all the data is contained in only 26 intensity levels (18–44).

markers, but the numerous other markers throughout the rest of the Bight area were not shown. This implied that their intensity did not fall within the range 29–31. There are two possible reasons. First, the dye markers deployed from the boats in the Hudson River Plume were filled with 100% fluoroscein dye in powdered form, whereas the helicopter-deployed tubes in the coastal waters were a 60–40 dye mix and may not have been as intense.

The other reason involves the satellite MSS resolution. The instantaneous field of view (IFOV) of the MSS is an area on the surface of the earth of about 200 × 200 ft (60 × 60 m). The dye streamers were only about 60 ft (18 m)

FIGURE 5.11. A close-up view of a dye marker.

wide, although they were sometimes many hundreds of feet long. Thus, dye affects less than one-third of any resolution element. The sensor integrates the signal caused by the light reflected from all of the area within the IFOV and records one intensity level for that resolution element. Thus, the recorded level is due to the reflected energy from the dye and from the background.

The background water near the dye implants in the Hudson River Discharge Plume was more sediment-laden and thus more highly reflective than the background waters near the dye markers in the coastal waters. The Hudson River Plume background water intensity range was 26–30; for the coastal waters this was 23–27. An analysis of these two factors suggested that the dye markers in the coastal waters would exhibit an intensity of 27.

An intensity slice was thus made for intensity levels 26–28. (The histogram of the scene window is shown in Figures 5.9 and 5.10. The results were positive; the dye markers in the coastal waters were enhanced. A close-up view of a dye marker located by this process is shown in Figure 5.11.

PREPARATION OF COLOR-CODED SCENE COMPONENT CLASSIFICATION PRODUCT

In order that all scene components could be viewed at once and still retain the component classification, each intensity slice Litton print was assigned a separate color code, and a composite display was generated.

Conclusion

The digital processing of the Landsat data succeeded in detecting the dye implant locations. These coincided with water sample collection points. The analysis of the water samples, their geometric relationship, and other concurrent data related to winds and tides gave rise to substantially improved understanding of the dispersion dynamics in the New York Bight.

NEW ORLEANS HYDROLOGY

Digital image processing allows operations to be custom tailored for a particular application. The Landsat image of the New Orleans area of Louisiana and Mississippi shown in Figure 5.12 is a standard rendering. The distribution of gray levels is chosen to be the best compromise among the diverse needs of the various application communities that seek to use this imagery. Figure 5.13 shows the same scene after a gray-scale adjustment has been digitally applied specifically for enhancing the subtle gray-scale variations in the Gulf of Mexico and Lake Pontchartrain areas of the picture. Of

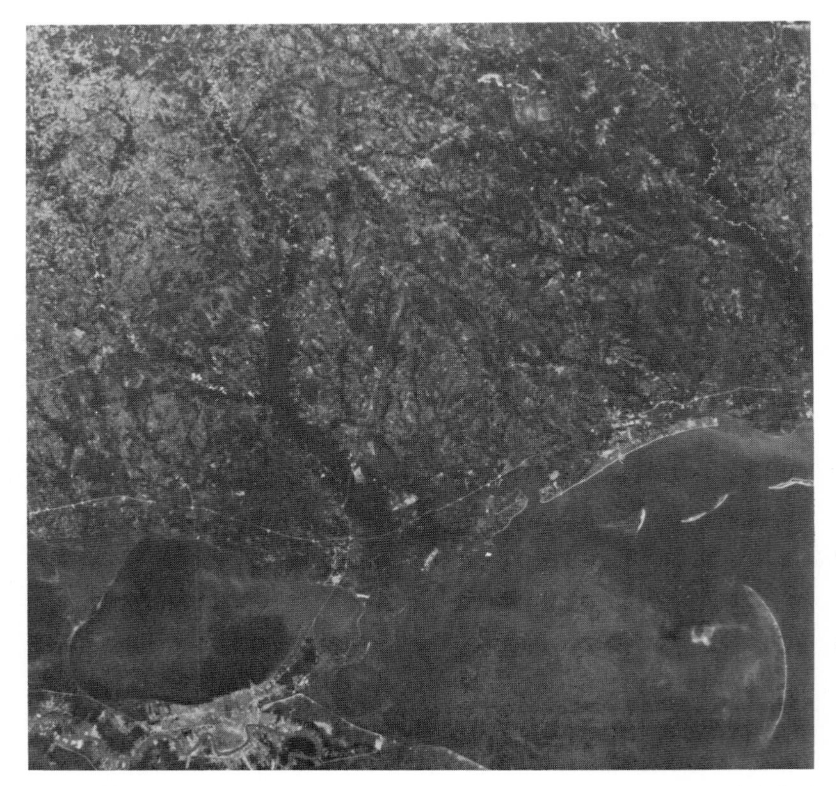

FIGURE 5.12. A Landsat image of the New Orleans area of Louisiana and Mississippi.

course, one pays for this particular improvement in hydrological interpretability with a loss of information in the land areas of the scene.

DIGITAL PROCESSING OF SATELLITE IMAGERY FOR GEOTHERMAL PROSPECTING[2]

Although many users of remote sensing data are familiar with the digital processing of Landsat imagery, less attention has been given by the user community to other sources of satellite acquired imagery and their applica-

[2]This section was originally presented to the American Society of Photogrammetry, Washington, D.C., March 1978, by R. M. Hord, R. Anderson, and M. C. Erskine, Jr.

FIGURE 5.13. The same scene as in Figure 5.12 after a gray-scale adjustment has been digitally applied.

tions. This section describes the processing of one such source, NOAA 3 imagery for geothermal prospecting.

Data

NOAA 3, one of the ITOS (Improved TIROS Operational Satellite) spacecraft, was launched into 1500-km (900-mile) sun-synchronous (quasi-polar) orbit with equator crossings near 9:00 a.m. (descending node) and 9:00 p.m. (ascending node), local sun time. The sensor housing was spin stabilized with a counter-rotating mass. The sensor housing maintained a constant Earth-pointing attitude by means of a closed-pitch detecting and correcting system and by roll and yaw magnetic torqueing commanded from the ground.

A Very High Resolution Radiometer (VHRR) described on page 31 was carried by NOAA 3. It has a scan rate of 400 rpm.

Each picture covers an area of approximately 2080 × 2080 km (1300 × 1300 miles), and two or three pictures are usually available per pass. The spacecraft and sensors are described in detail in the *Catalog of Operational Satellite Products* and in Schwalb (1972).

The orbit is designed with 78° inclination and a 115-min period to cover the entire globe in 12 hr. The temperature range, for emissivity = 1, is from 310°

FIGURE 5.14. A gray-scale adjusted 1500 × 1500 window from a digital image that includes California, Nevada, and other western states.

K (black) to 180° K (white), a scale arranged to display cold areas as bright and hot areas as dark so clouds appear white (Yates & Bandeen, 1975). The particular image employed in this study was provided by the Environmental Products Group, National Environmental Satellite Service, in August 1974. The image was acquired on 1 April 1974.

The user, then, is provided with a digital image with 5340 pixels (8 bits per pixel) on each line, and up to 6900 lines per frame. Figure 5.14 shows a gray-scale adjusted 1500 × 1500 window from this image that includes California, Nevada, and other western states.

Geothermal Considerations

Economic geothermal energy sites have assumed a growing importance in recent years. Although geothermal steam utilization began on an industrial scale in Italy in 1904, it was not until 1960 that a dry steam reservoir was used on a commercial scale in the United States (The Geysers, California) for the production of electricity. Since then, exploration and development have been concentrated in 11 western states where volcanic activity and faulting led to the identification of a number of high temperature areas. A published map of Known Geothermal Resource Areas (KGRA) is shown in Figure 5.15.

Reconnaissance geothermal prospecting is becoming more sophisticated. For example, some groups are using aeromagnetic data. Processing aeromagnetic data allows, among other interpretations, estimation of the Curie point depth. The Curie point depth is defined as the depth at which the temperature is sufficiently high (840° K) that magnetite no longer can retain its magnetic properties. In practice, a temperature of 775° K is generally used, since fall-off of magnetization is rapid beyond that point. Usually the Curie point depth is between 10 km and 30 km (6 and 18 miles); when it is 6 km (3.6 miles) or less, the location may be considered a geothermal prospect. Since temperatures of 475° K are useful as an energy source, a 3-km (1.8-mile) Curie point depth suggests that exploitable thermal energy may be encountered at about 1250 m (4000 ft).

Since aeromagnetic surveys can be expensive, other indicators are sought. One in common use is aerial thermal infrared imagery, which permits direct observation of high surface temperatures and avoids wasted effort in the calculation of the Curie point. Another example of a surface indicator is anomalous heat flow, as in the Battle Mountain Heat Flow High, just south of Winnemucca, Nevada. Such areas are assessed in terms of heat flow units (1 HFU = 1 μcal flowing through 1 cm^2 in 1 sec). The Battle Mountain high has

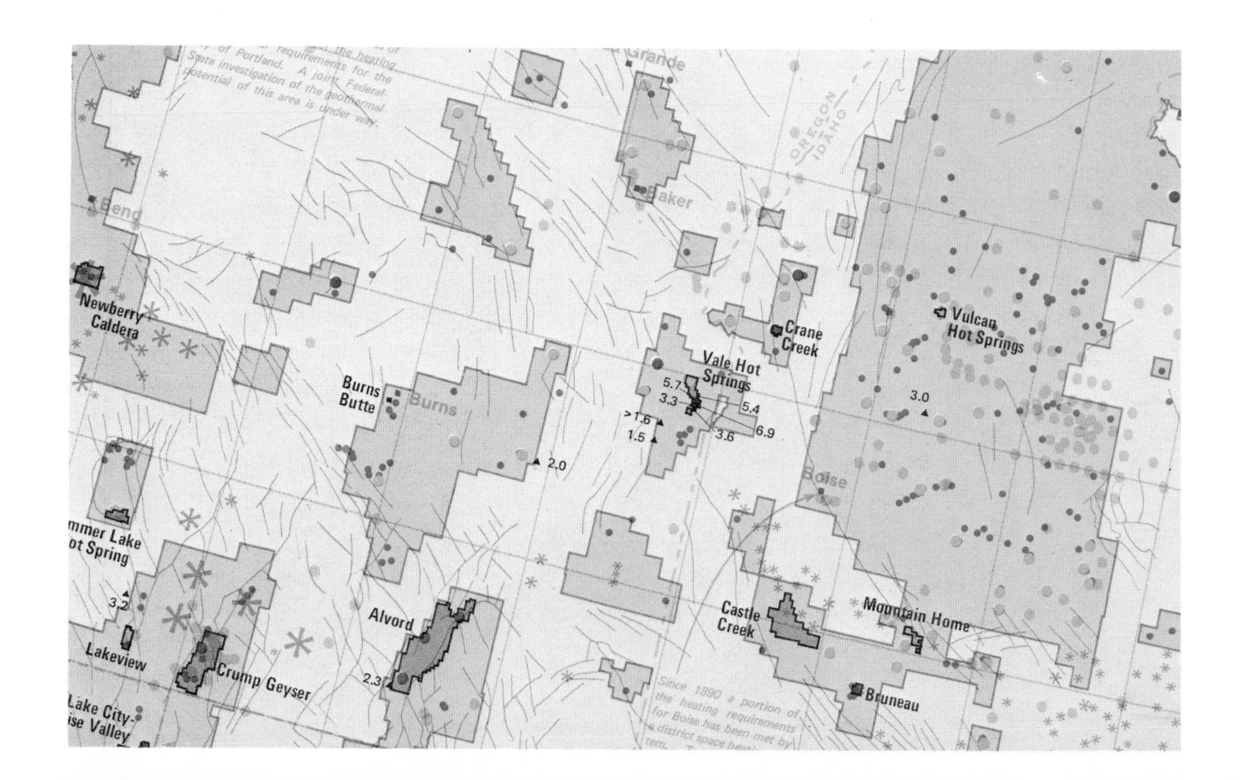

Newberry Caldera
Bend
Burns Butte
Burns
2.0
Vale Hot Springs
5.7
3.3
>1.6
1.5
5.4
3.6
6.9
Crane Creek
Baker
OREGON
IDAHO
Vulcan Hot Springs
3.0
Boise
Castle Creek
Mountain Home
Bruneau
Alvord
2.3
Crump Geyser
Lakeview
3.2

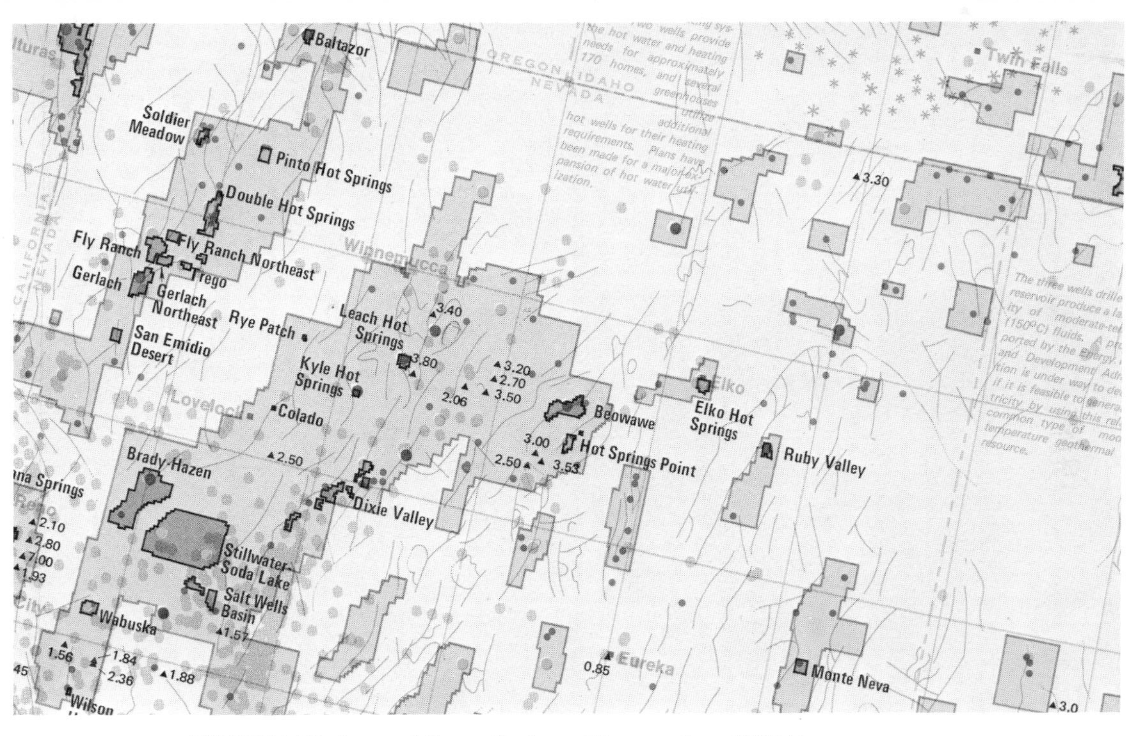

FIGURE 5.15. A map of Known Geothermal Resource Areas (KGRA).

been measured at 3 HFU, whereas the surrounding area is characterized by 1.5 HFU. The average over the earth's surface is 1.0 HFU.

Interpretation of such surface indicators as aerial infrared imagery requires substantial skill. Looking for "hot spots" 2° or 3° K different from the background in the context of a 30° K daily temperature change, variations in albedo due to vegetation of thermal inertial, can be a challenge.

Processing

The goal of the effort was to determine whether satellite thermal infrared imagery with computer processing would afford surface indications of geothermal prospects (Williams & Friedman, 1970). A blind experiment using computer processing was conducted in an attempt to isolate hot spots, after which a comparison was made with a map of KGRA. Thorough processing was not attempted; rather, a feasibility test was undertaken.

North–south temperature gradient compensation was performed. It was observed, not surprisingly, that northern locales registered somewhat cooler temperatures than southern locales. Accordingly, the average gray level for each image line was calculated. These averages, as a function of line number, were subjected to linear fitting, and a gray-scale adjustment was applied to cloud-free image strips on a line by line basis. After adjustment, all lines had approximately the same average gray level. No east–west temperature gradient compensation was performed, although solar illumination effects would indicate a possible benefit from such processing.

The gray levels were also adjusted to take full advantage of the dynamic range of the display unit (a hardcopy image recorder). At this point, each increment in gray level corresponded to a temperature difference of .5° K. The final processing step was thresholding. Pixels hotter than a temperature represented by a gray-level value T are shown as black; all others are white. Figures 5.16–5.18 show the image for $T = 147$, $T = 159$, and $T = 168$. Figure 5.17 was used for interpretation. An abundance of isolated hot spots exhibiting structure in their distribution were observed.

Interpretation

Among the recognized hot spots are the Battle Mountain Heat Flow Anomaly, the Castle Creek KGRA, Vale, Oregon Hot Springs KGRA; Fly Ranch Hot Springs, Double Hot Springs, the Basalt of Shaffer Mountain, Surprise Valley–Lake City KGRA, Black Andesites at Sutter Buttes and the Thermal Springs areas of the north edge of the Snake River plain. The solar retention

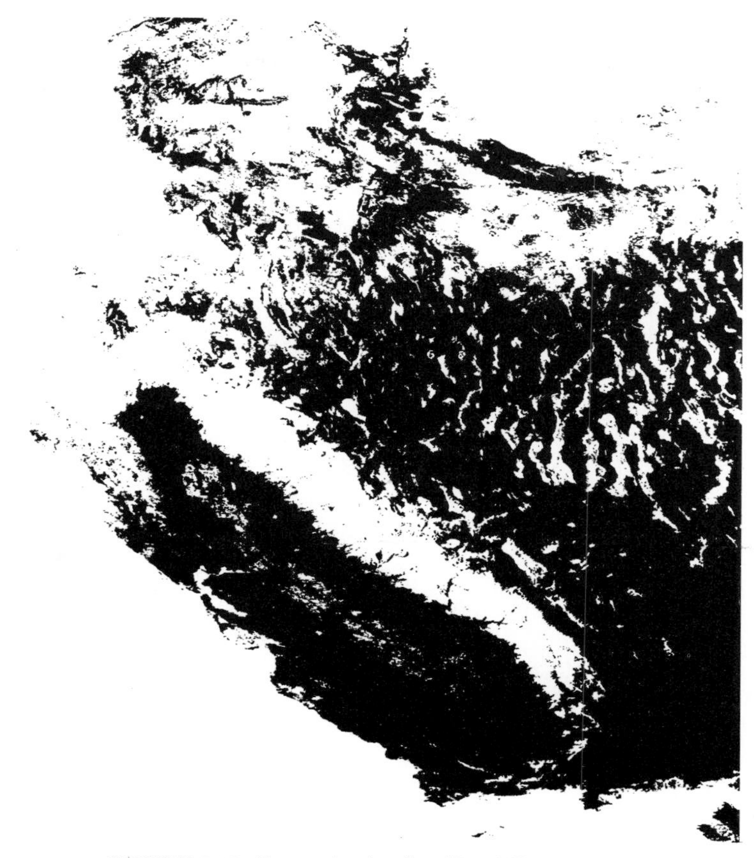

FIGURE 5.16. Energy density slice (T = 147).

and nighttime reradiation of energy from outcropping black rock represents an interpretation noise problem also common to airborne scanner surveys. It seems to be less of a problem in the satellite image.

Discussion

One of the advantages of VHRR imagery for this application is the relatively large size of the ground-resolved area, about 160 acres (65 ha). The corresponding gray-level represents an average temperature over this area, so that small hot spots, like small hot springs, which generally do not indicate a geothermal prospect, are integrated away. Because a simple feasibility study

FIGURE 5.17. Energy density slice ($T = 159$).

was intended, geometric correction to a Universal Transverse Mercator projection was not performed. In retrospect, geometric correction to reduce distortions would have expedited interpretation.

The method described here, even if performed thoroughly, is unlikely to be of much practical value in a well-explored geographic area. What is suggested, rather, is that this technique may be of practical value in planning aeromagnetic and other surveys of unexplored or even partially explored areas.

WASHINGTON, D.C. LAND USE MAPPING

The Geography Program of the U.S. Geological Survey, Reston, Virginia, used Landsat imagery and the LARSYS program implemented on the ad-

FIGURE 5.18. Energy density slice (T = 168).

vanced parallel processing Illiac IV computer to produce the land cover map of Washington, D.C. shown in Figure 5.19. For a land use change study, see Gordon (1980).

CROP ACREAGE ESTIMATION

Because of the very rapid changes in the condition of agricultural crops and the influence of crop yield predictions on the world market, the need for accurate, timely information is particularly acute in agricultural information systems. It is for these reasons that, as remote sensing technology had developed, the potentials for using this technology have received widespread attention (MacDonald & Hall, 1980).

Crop acreage estimation consists of two parts, the mensuration of field sizes and the categorization of those fields by crop type. The mensuration process can sometimes be facilitated by manipulating imagery to make field boundaries more distinct. Figure 5.20 shows part of a Landsat image obtained over Divide County, North Dakota, where the ratio of band 5:band 6 has been generated since this rendering exhibits the field boundaries more clearly than either band 5 or band 6 alone. The categorization of those fields by crop type is performed by multispectral classification, as discussed earlier (see Chapter 4, page 96).

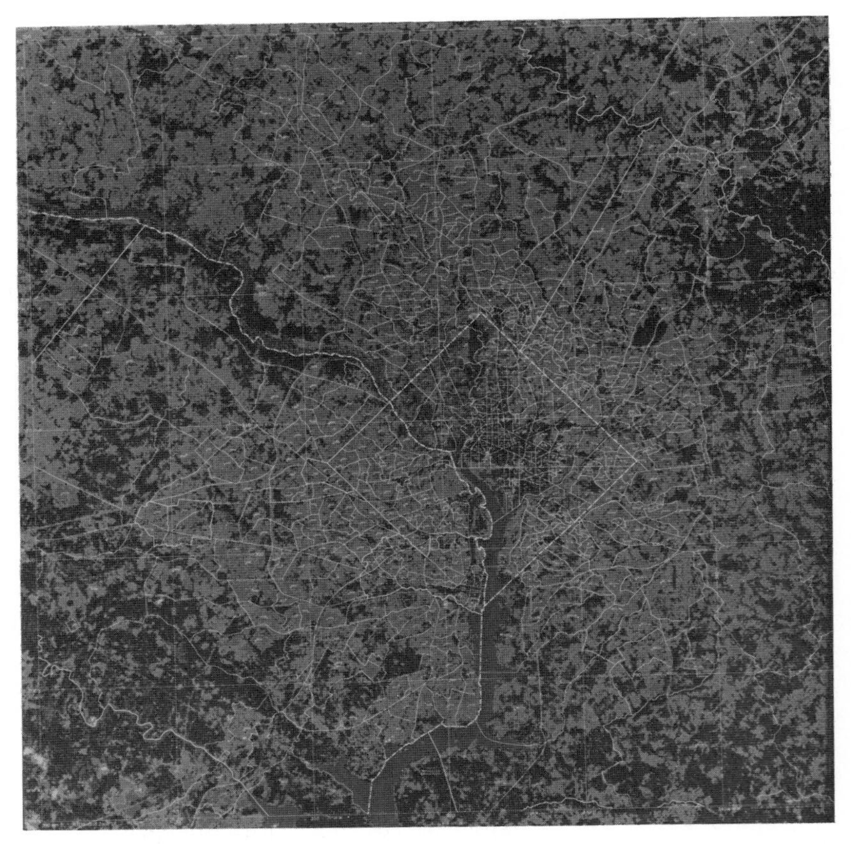

FIGURE 5.19. Land cover map of Washington, D.C. produced using Landsat imagery and the LARSYS program.

The United States, government is engaged in a large-crop estimation activity because the economic benefits to be obtained are massive. A November 1978 report in *Aviation Week and Space Technology* magizine reads, in part:

> The three-year large area crop inventory experiment (Lacie), using Landsat multispectral scanner-generated imagery, has proved the feasibility of global monitoring of food and fiber production by satellite and paved the way for systems that would be fully operational in 1980–1984.
>
> The U.S. Department of Agriculture is making the transition from being a partner with National Aeronautics and Space Administration and the National Oceanic and Atmospheric Administration in the recently concluded Lacie proof-of-concept program

FIGURE 5.20. Part of a Landsat image obtained over Divide County, North Dakota. The ratio of band 5 to band 6 has been generated.

to developing Lacie as an operational test system geared to support new initiatives in crop forecasting.

The Agriculture Department will continue as a partner with NASA and NOAA in the research and development of a second-generation satellite crop assessment system with expanded capabilities. The new research program is oriented toward using Landsat–D, the fourth remote-sensing satellite that is planned to carry an improved multispectral sensor with two to three times the resolution of Landsat–3.

Where Lacie's mission was to prove the feasibility of using Landsat for yield assessment of one crop, wheat, the next Landsat launched will be used to determine the feasibility of satellite assessment of eight crops on a global scale.

With the U.S. becoming an increasing exporter of key foods, it has been estimated by Lacie participants that the economic benefits to this country's agricultural industry from improved estimates of foreign crop yields could be on the order of $300 million annually, with wheat alone accounting for $200 million.

The present reliability of the [Dept. of Agriculture] system of developing foreign estimates from a variety of sources has been low because many major foreign producers lack accurate data sources.

> The Lacie experiment highlighted the potential impact that a credible satellite crop yield assessment system could have on world food marketing, Administration policy, transportation and other related factors.
>
> In 1977, during Phase 3 of Lacie, it was decided to test the accuracy of Soviet wheat crop yield data by using Landsat–3 to assess the USSR's total production from early season through harvest.
>
> According to Lacie experimenters, in January 1977, the Soviet Union officially announced it expected a total grain crop of 213.3 million metric tons. This was about 13% higher than the country's 1971–1976 average.
>
> Since Soviet wheat historically has accounted for 48% of its total grain production, the anticipated wheat yield would have been about 102 million metric tons for the year.
>
> In late January, 1978, the USSR announced that its 1977 wheat production had been 92 million metric tons.
>
> Lacie computations, made after the Soviet harvests, but prior to the USSR release of figures, estimated Russian wheat production at 91.4 million metric tons. The [Dept. of Agriculture] final estimate was 90 million metric tons.
>
> Lacie experimenters said the Agriculture Dept. assessments of Soviet wheat yields have had an accuracy of only 65/90. This means that the department's conventionally collected data could have an accuracy within ± 10% only 65% of the time. The Lacie program was designed to provide a crop yield assessment accuracy of 90/90, or within ± 10% in 90% of the years the system is used [pp. 65].

Additional information about global crop forecasting can be found in MacDonald and Hall (1980).

OCEAN STATE

On occasion, even the ocean surface state can be observed with suitable processing of satellite imagery. SEASAT data is specifically designed for this. VHRR imagery (page 148 of this chapter and page 31 of Chapter 2) can be processed to provide a temperature contours of the coastal waters (see also Emery & Mysak, 1980). In Figure 5.21 of this section, Landsat scene 1118–07021 band 7 has been gray-scale adjusted to bring out ocean wave patterns. An interference wave pattern can be seen moving toward the coast, having been produced by the apertures in the barricade provided by the string of islands off shore.

URBAN PLANNING

Effective urban and regional planning depends on accurate information regarding current land usage. Enhanced satellite imagery can contribute sub-

FIGURE 5.21. A Landsat scene that has been gray-scale adjusted to bring out ocean wave patterns. An interference wave pattern can be seen moving to the coast.

stantially to this information base. For example, Figure 5.22 shows the streets of Tehran, Iran, Laplacian-enhanced from Landsat imagery. Urban population monitoring is also discussed in Welsh (1980).

HIGH TIDE–LOW TIDE

Laws protecting the United States wetlands define those areas in terms of tidal extrema. Mapping those from the ground is time consuming and labor

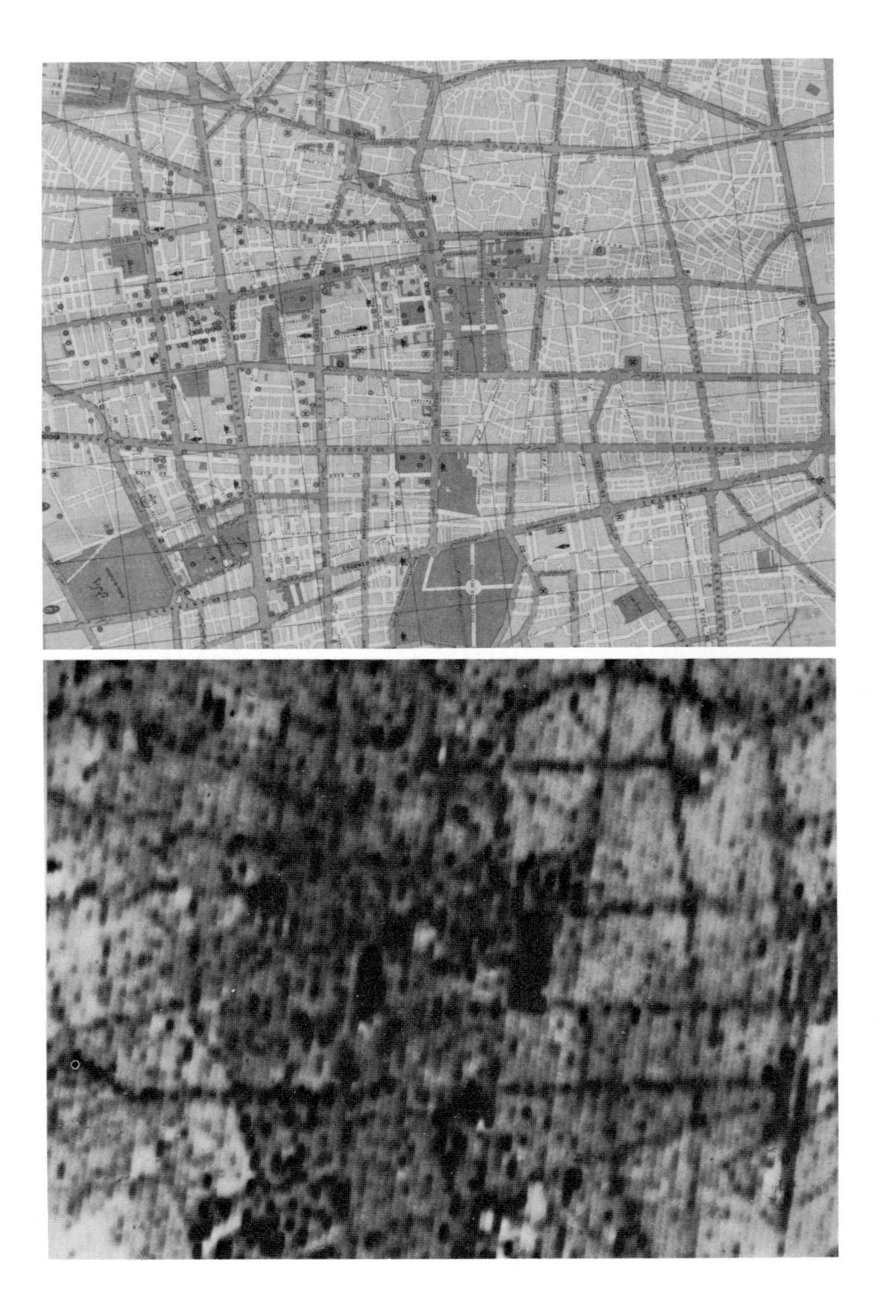

Earth Resources Technology Satellite imagery has been demonstrated as having significant applications to urban and regional land use planning, including monitoring of land use change, land use mapping and land use suitability analysis. ERTS also effectively serves as a graphic communications media, important in both the land planning processes and resource policy decision making.

Urban and regional planning and land use management processes are, by nature, iterative. ERTS data, because of its scale, resolution and repetitive nature can be effectively applied to monitoring the impact of regional planning programs and to evaluating the results.

ERTS data provides a synoptic view from which regional information concerning land use and economic activities may be analyzed. The detail of landscape information available from ERTS is limited by scale and resolution of the images. Urban areas often contain highly structured and complex environments, that cannot visually be analyzed using standard photographic products. To improve the quality and application of ERTS photoproducts, a variety of specialized computer processing and Digital Image Processing techniques can be applied for specific objectives. An example is an ERTS image of Tehran and vicinity, acquired October 5, 1972. The image was digitally processed to sharpen transportation networks and provide analysis of land use within the urban areas.

Comparison with a 1970 map shows recent change in the urban environment. Major new building developments may be readily identified to update general land use maps. In the computer processed image, major city thoroughfares are easily detected. Major parks are separable from large administrative building complexes in the central portion of Tehran. Sample urban sectors, indicated by yellow outlines on the city map, can be precisely located and further analyzed by computer processing of data acquired from other sources and supporting aircraft photography.

FIGURE 5.22. The streets of Tehran, Iran Laplacian-enhanced from Landsat imagery.

FIGURE 5.23. High tide at Little Ferry, N.J.

intensive. Enhanced satellite imagery can provide an effective alternative approach. Figures 5.23 and 5.24 show a High tide–low tide comparison.

SPACE OBJECTS

Many space missions carry image sensors and telemetry systems that relay the imagery to earth. America's space images, that is, pictures of Mars, the Moon, and such are generally reconstructed at the Jet Propulsion Laboratory,

FIGURE 5.24. Low tide at Little Ferry, N.J.

Pasadena, California. Noise removal and contrast adjustment are performed in a routine fashion. Figure 5.25 shows an enhanced view of Phobos, the 20.8-km (13-mile) diameter Martian moon. Figure 5.26, on the other hand, shows an Earth-orbiting vehicle, the satellite Pegasus, as enhanced from imagery acquired by a ground-based telescope.

The Pegasus 1 spacecraft assembly, launched by NASA in 1965, reentered the Earth's atmosphere in September 1978. It was used to gather micrometeoroid data for use in the design of spacecraft. Data collection stopped in January 1968. Pegasus 1 was one of three such spacecraft. Pegasus 2 is still in

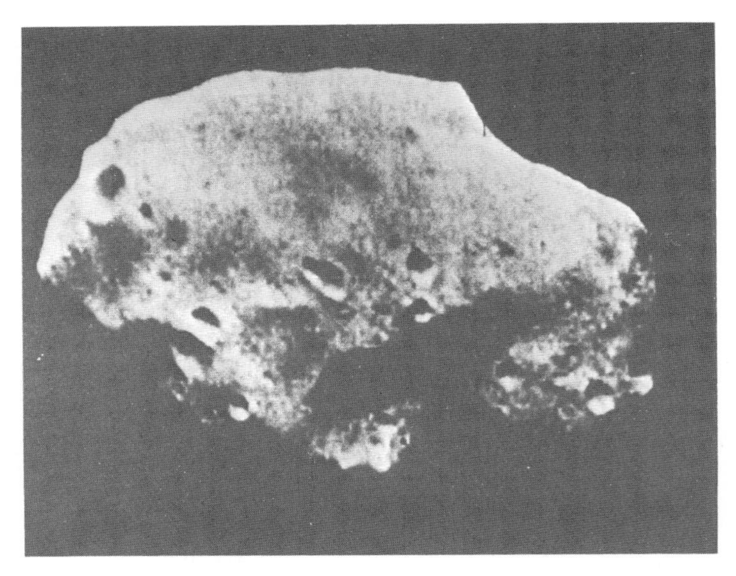

FIGURE 5.25. An enhanced view of Phobos, the innermost moon of Mars.

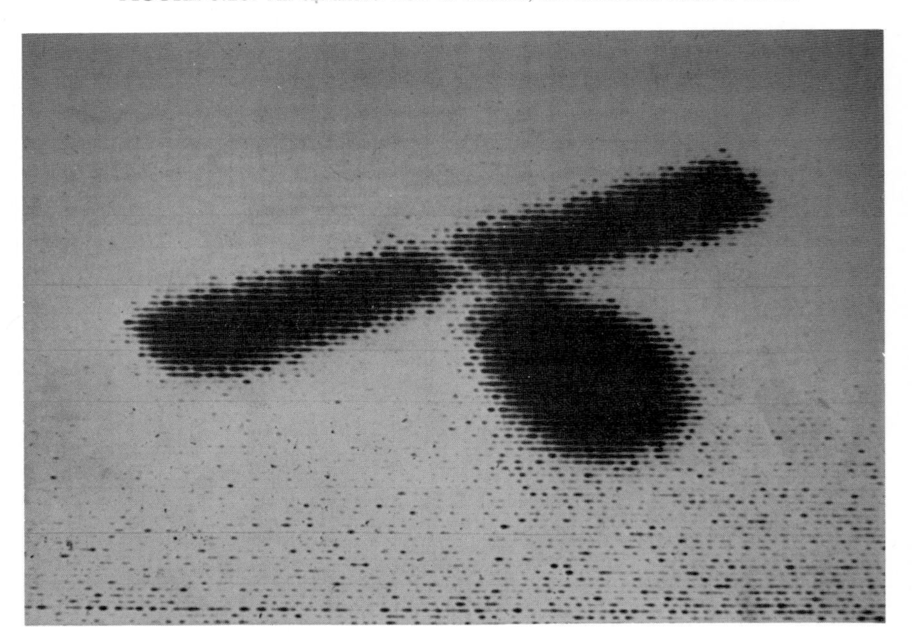

FIGURE 5.26. The earth-orbiting satellite Pegasus, as enhanced from imagery acquired by a ground-based telescope.

orbit, and Pegasus 3 had reentered previously. Pegasus 1 weighed 10,350 Kg (23,000 lb) and was 21 m (70 feet) long. The "wing span" was 28.8 m (96 ft). It had an 87-min period.

REFERENCES

Bulban, E. J. Landsat crop assessments expanding. *Aviation Week and Space Technology, 13*(Nov. 1978), 109(20), 65.

Emery, & Mysak. Dynamical interpretation of satellite-sensed thermal features off Vancouver Island, *Journal of Physical Oceanography,* 1980, *10,* 961–970.

Gordon, S. I. Utilizing Landsat imagery to monitor land use change: A case study in Ohio. *Remote Sensing of Environment,* 1980, *9*(3) 189–196.

Johnson, R. W. Remote sensing and spectral analysis of plumes from ocean dumping in the New York Bight apex. *Remote Sensing of* Environment, 1980, *9*(3), 197–209.

MacDonald, R. B., & Hall, F. G. Global crop forecasting. *Science,* 1980, *208,* 670–679.

Macomber, R. T., Mairs, R., & Stanczuk, D. *A study of the surface fluid flow characteristics of the New York Bight and the development of Research data display techniques.* Washington, D.C.: Earth Satellite Co., 1974.

National Oceanic and Atmospheric Administration (NOAA). *Catalog of operational satellite products* (Technical Memorandum NSS 53). Washington, D.C.: U.S. Department of Commerce, 1974.

Schwalb, A. *Modified version of the improved TIROS operational satellite (ITOS D-G)* (NOAA Technical Memorandum NESS 35). Washington, D.C.: U.S. Department of Commerce, 1972.

Welsh, R. Monitoring urban population and energy utilization patterns from satellite data. *Remote Sensing of Environment,* 1980, *9*(1), 1–9.

Yates, H. W., & Bandeen, W. R. Meteorological applications of remote sensing from satellites. *Proceedings of the IEE,* 1975, *63,* 148.

Research Topics

Although it is useful to know what digital image processing can do and how it is done, it is also useful to know the limits of this technology. Many problems persist unsolved. One way to gain an appreciation of the state of the art is to review efforts to move this knowledge boundary forward. This chapter gives in-depth accounts of three of these research activities.

This approach is chosen to show how a number of techniques can be drawn together to form a processing sequence that addresses a specific task. Unfortunately, this approach limits the breadth of topics that can be covered. The range of research topics in the area of digital image processing of remotely sensed data is quite diverse, as a review of the related reading listed at the end of this volume will confirm.

Some of the other current research topics in the field include change detection, partial pixel classification, apparent thermal inertia determination, crop yield estimation, relaxation labeling, signature extension, atmospheric correction, and a host of others. Naturally, these technique development efforts are accompanied by research to improve image storage and retrieval methods, to speed up processing time, and to reduce processing costs; there are efforts both to reduce human involvement to permit on-board processing and to increase human involvement so that the uniquely human skills can be more effectively employed. In general, the field exhibits a degree of research vitality that well may be envied by scholars of some other disciplines.

DIGITAL IMAGE SHAPE DETECTION[1]

The ability to determine algorithmically the presence of specified objects in test imagery is desirable in many circumstances. For example, looking for parked airplanes or road intersections are common undertakings. Typically, a reference image containing an example of the object of interest is compared with a test image. The four primary characteristics of the object used for this comparison are:

1. *Translation:* location of the object's centroid in the format
2. *Orientation:* rotation of the object's principal axes with respect to the format axes
3. *Size:* scale of the object with respect to the sample spacing
4. *Shape:* the distribution of gray-level values of the object

Often, only the shape is relevant, that is, a match is sought between the test image and the reference image even if they differ in translation, orientation, and scale. It is desirable in these cases to transform the image pair in ways that are independent of these irrelevant geometric features. Shape detection has been a topic of interest to the digital image pattern recognition community for many years (Hawkins *et al.*, 1966), and it remains a prominent topic of discussion in the literature (Agrawala & Kulkarmi, 1977; Danielsson, 1978; Dinstein & Silberberg, 1980; Wong & Hall, 1978). Two survey articles (Pavlidis, 1978; Rosenfeld, 1978) attest to the popularity of the topic.

The *power spectrum* of an image is well known to be independent of translation. The Mellin transform has been shown to be scale independent (Huang, Russell, & Chen, 1975). The Polar–Cartesian (POLCAR) Transform described below, converts rotation into translation (Hord & Sheffield, 1975). The power spectrum of the POLCAR transform is then independent of rotation. Hence a combination of these procedures, performed successively, will allow shape to be matched to shape, independent of translation, rotation, and size. Mention of this approach is found in the literature (Brousil & Smith, 1967; Rosenfeld & Kak, 1976); this chapter explores this approach in more detail.

[1]This section originally published as Hord, R. M., Digital image shape detection, *Proceedings of the National Computer Conference*, 1979, *48*, 7, 243.

POLCAR Transform (32 × 32)

In a $N \times N$ image, let I be the row index and J be the column index with $(I,J) = (1,1)$ in the upper left-hand corner. Let

$$S = (N/2) + 1.01 \quad (1)$$
$$X = J - S \quad (2)$$
$$Y = I + S \quad (3)$$
$$R = \text{SQRT}\ (X \times X + Y \times Y) + .1 \quad (4)$$
$$T = \text{ARCTAN}(Y/X) + 3.14159 \quad (5)$$

Then for $N = 32$, the x axis is origined at 17.01 and is positive to the right and the y axis is origined at 17.01 and is positive upward in the I,J system. R ranges, for integer I,J, between .11 and 22.74, whereas T ranges between 0 and 6.2832. Let (in FORTRAN notation)

```
IR = INT (LOG(R) * 13.155 + 14.2)        (6)

ITH = INT (T * 5. + 1.)                  (7)
```

Then *IR* is an index ranging between 1 and 32 and *ITH* is an index ranging between 1 and 32. Let *IR* be a row index and *ITH* be a column index with (*IR*, *ITH*) = (1,1) in the upper-left corner of a 32 × 32 array. This array is termed the POLCAR transform of the input array: $G(IR, ITH) = G(I,J)$.

An input array of horizontal lines is shown in Figure 6.1; the geometry of the POLCAR transform of Figure 6.1 is shown in Figure 6.2. Figure 6.3 shows the isolines of Figure 6.2. Figure 6.4 shows the full POLCAR transform of Figure 6.1, including gray-level amplification as a function of R^2. Asterisks indicate format overflow for values greater than 99.

A set of vertical lines are shown in Figure 6.5; the geometric features of the POLCAR transform of Figure 6.5 are shown in Figure 6.6 with the isolines drawn in in Figure 6.7. Note that Figure 6.7 resembles Figure 6.3, shifted eight spaces to the right, corresponding to a 90° rotation of the input array. Figure 6.8 shows the array of *IR* values as a function of I,J; Figure 6.9 shows the isolines of Figure 6.8. The array of *ITH* values as a function of I,J is shown in Figure 6.10. The isolines of Figure 6.10 are shown in Figure 6.11.

A FORTRAN subroutine implementing the POLCAR Transform is given in Table 6.1. It is intended that the input array, IMAG, is the power spectrum of a 32 × 32 image, rearranged to emulate an optical power spectrum so that the zero-frequency component is in position at (17,17). By setting the origin at (17.01, 17.01), we know that R is strictly positive.

```
1 1 1 1 1 1 1 1 1 1 1 1 1 1 1 1 1 1 1 1 1 1 1 1 1 1 1 1 1 1 1 1
2 2 2 2 2 2 2 2 2 2 2 2 2 2 2 2 2 2 2 2 2 2 2 2 2 2 2 2 2 2 2 2
3 3 3 3 3 3 3 3 3 3 3 3 3 3 3 3 3 3 3 3 3 3 3 3 3 3 3 3 3 3 3 3
4 4 4 4 4 4 4 4 4 4 4 4 4 4 4 4 4 4 4 4 4 4 4 4 4 4 4 4 4 4 4 4
5 5 5 5 5 5 5 5 5 5 5 5 5 5 5 5 5 5 5 5 5 5 5 5 5 5 5 5 5 5 5 5
6 6 6 6 6 6 6 6 6 6 6 6 6 6 6 6 6 6 6 6 6 6 6 6 6 6 6 6 6 6 6 6
7 7 7 7 7 7 7 7 7 7 7 7 7 7 7 7 7 7 7 7 7 7 7 7 7 7 7 7 7 7 7 7
8 8 8 8 8 8 8 8 8 8 8 8 8 8 8 8 8 8 8 8 8 8 8 8 8 8 8 8 8 8 8 8
9 9 9 9 9 9 9 9 9 9 9 9 9 9 9 9 9 9 9 9 9 9 9 9 9 9 9 9 9 9 9 9
1010101010101010101010101010101010101010101010101010101010101010
1111111111111111111111111111111111111111111111111111111111111111
1212121212121212121212121212121212121212121212121212121212121212
1313131313131313131313131313131313131313131313131313131313131313
1414141414141414141414141414141414141414141414141414141414141414
1515151515151515151515151515151515151515151515151515151515151515
1616161616161616161616161616161616161616161616161616161616161616
1717171717171717171717171717171717171717171717171717171717171717
1818181818181818181818181818181818181818181818181818181818181818
1919191919191919191919191919191919191919191919191919191919191919
2020202020202020202020202020202020202020202020202020202020202020
2121212121212121212121212121212121212121212121212121212121212121
2222222222222222222222222222222222222222222222222222222222222222
2323232323232323232323232323232323232323232323232323232323232323
2424242424242424242424242424242424242424242424242424242424242424
2525252525252525252525252525252525252525252525252525252525252525
2626262626262626262626262626262626262626262626262626262626262626
2727272727272727272727272727272727272727272727272727272727272727
2828282828282828282828282828282828282828282828282828282828282828
2929292929292929292929292929292929292929292929292929292929292929
3030303030303030303030303030303030303030303030303030303030303030
3131313131313131313131313131313131313131313131313131313131313131
3232323232323232323232323232323232323232323232323232323232323232
```

FIGURE 6.1. Horizontal lines.

```
0 0 0 0 0 0 0 0 0 0 0 0 0 0 0 0 0 0 0 0 0 0 0 0 0 0 017 0 0 0 0
0 0 0 0 0 0 0 0 0 0 0 0 0 0 0 0 0 0 0 0 0 0 0 0 0 0 0 0 0 0 0 0
0 0 0 0 0 0 0 0 0 0 0 0 0 0 0 0 0 0 0 0 0 0 0 0 0 0 0 0 0 0 0 0
0 0 0 0 0 0 0 0 0 0 0 0 0 0 0 0 0 0 0 0 0 0 0 0 0 0 0 0 0 0 0 0
0 0 0 0 0 0 0 0 0 0 0 0 0 0 0 0 0 0 0 0 0 0 0 0 0 0 0 0 0 0 0 0
0 0 0 0 0 0 0 0 0 0 0 0 0 0 0 0 0 0 0 0 0 0 0 0 0 0 0 0 0 0 0 0
0 0 0 0 0 0 0 0 0 0 0 0 0 0 0 0 0 0 0 0 0 0 0 0 0 0 0 0 0 0 0 0
0 0 0 0 0 0 0 0 0 0 0 0 0 0 0 0 0 0 0 0 0 0 0 0 0 0 0 0 0 0 0 0
0 0 0 0 0 0 0 0 0 0 0 0 0 0 0 0 0 0 0 0 0 0 0 0 0 0 0 0 0 0 0 0
0 0 0 0 0 0 0 0 0 0 0 0 0 0 0 0 0 0 0 0 0 0 0 0 0 0 0 0 0 0 0 0
0 0 0 0 0 0 0 0 0 0 0 0 0 0 0 0 0 0 0 0 0 0 0 0 0 0 0 0 0 0 0 0
0 0 0 0 0 0 0 0 0 0 0 0 0 0 0 0 0 0 0 0 0 0 0 0 0 0 0 0 0 0 0 0
0 0 0 0 0 0 0 0 0 0 0 0 0 0 0 0 0 0 0 0 0 0 0 0 0 0 0 0 0 0 0 0
0 0 0 0 0 0 018 0 0 0 0 0 0 017 0 0 0 0 0 0 016 0 0 0 0 0 0 017
0 0 0 0 0 0 0 0 0 0 0 0 0 0 0 0 0 0 0 0 0 0 0 0 0 0 0 0 0 0 0 0
0 0 018 0 0 0 0 0 0 018 0 0 0 0 0 0 016 0 0 0 0 0 0 016 0 0 0 0
0 0 0 0 0 0 0 0 0 0 0 0 0 0 0 0 0 0 0 0 0 0 0 0 0 0 0 0 0 0 0 0
0 0 0 0 0 0 019 0 0 0 0 0 0 017 0 0 0 0 0 0 015 0 0 0 0 0 0 017
0 018 0 019 0 0 0 019 0 018 0 0 0 016 0 015 0 0 015 0 0 016 0 0
018 019 0 02020 020 019 0 01817 016 015 014 014 014 015 016 017
0 019 020 0 0 0 0 020 019 0 0 0 0 015 014 0 0 0 0 014 015 0 0 0
0181920 0212121 0212120 019181716 01514 01313131313 014 0151617
1819 020212222222222 021201918171615l4 013121212121213 014151617
181920212222232323232222212019171615141312111111111111213141516l7
18192122232424242424232221201918161514121111101010111112131416l7
182021232425262626252524232119181614131211 9 9 9 9 9101213141617
19212324262727282827262524222018161412l0 9 8 7 7 7 8 91012141617
192224262729293030292827252321181614121O 8 7 6 5 6 7 8 911131617
2022252829313232323231292724211916131l 8 6 5 4 3 4 5 6 810131517
2023273032323 2 0 0323232292622 0 012 9 6 4 2 1 1 1 2 4 6 9121517
0 0273232 0 0 0 0 0 03230 0 0 0 0 0 0 5 2 0 0 0 0 0 2 4 7 0 0 0
0 0 0 0 0 0 0 0 0 0 0 0 0 0 0 0 0 0 0 0 0 0 0 0 0 0 0 1 0 0 0 0
```

FIGURE 6.2. POLCAR transform of horizontal lines.

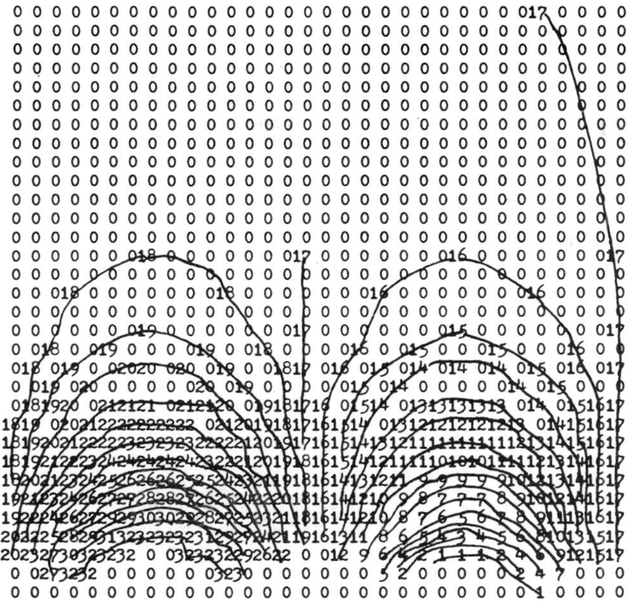

FIGURE 6.3. Isolines of POLCAR transform of horizontal lines.

FIGURE 6.4. Edge-enhanced POLCAR transform of horizontal lines.

```
1 2 3 4 5 6 7 8 91011121314151617181920212223242526272829303132
1 2 3 4 5 6 7 8 91011121314151617181920212223242526272829303132
1 2 3 4 5 6 7 8 91011121314151617181920212223242526272829303132
1 2 3 4 5 6 7 8 91011121314151617181920212223242526272829303132
1 2 3 4 5 6 7 8 91011121314151617181920212223242526272829303132
1 2 3 4 5 6 7 8 91011121314151617181920212223242526272829303132
1 2 3 4 5 6 7 8 91011121314151617181920212223242526272829303132
1 2 3 4 5 6 7 8 91011121314151617181920212223242526272829303132
1 2 3 4 5 6 7 8 91011121314151617181920212223242526272829303132
1 2 3 4 5 6 7 8 91011121314151617181920212223242526272829303132
1 2 3 4 5 6 7 8 91011121314151617181920212223242526272829303132
1 2 3 4 5 6 7 8 91011121314151617181920212223242526272829303132
1 2 3 4 5 6 7 8 91011121314151617181920212223242526272829303132
1 2 3 4 5 6 7 8 91011121314151617181920212223242526272829303132
1 2 3 4 5 6 7 8 91011121314151617181920212223242526272829303132
1 2 3 4 5 6 7 8 91011121314151617181920212223242526272829303132
1 2 3 4 5 6 7 8 91011121314151617181920212223242526272829303132
1 2 3 4 5 6 7 8 91011121314151617181920212223242526272829303132
1 2 3 4 5 6 7 8 91011121314151617181920212223242526272829303132
1 2 3 4 5 6 7 8 91011121314151617181920212223242526272829303132
1 2 3 4 5 6 7 8 91011121314151617181920212223242526272829303132
1 2 3 4 5 6 7 8 91011121314151617181920212223242526272829303132
1 2 3 4 5 6 7 8 91011121314151617181920212223242526272829303132
1 2 3 4 5 6 7 8 91011121314151617181920212223242526272829303132
1 2 3 4 5 6 7 8 91011121314151617181920212223242526272829303132
1 2 3 4 5 6 7 8 91011121314151617181920212223242526272829303132
1 2 3 4 5 6 7 8 91011121314151617181920212223242526272829303132
1 2 3 4 5 6 7 8 91011121314151617181920212223242526272829303132
1 2 3 4 5 6 7 8 91011121314151617181920212223242526272829303132
1 2 3 4 5 6 7 8 91011121314151617181920212223242526272829303132
1 2 3 4 5 6 7 8 91011121314151617181920212223242526272829303132
1 2 3 4 5 6 7 8 91011121314151617181920212223242526272829303132
```

FIGURE 6.5. Vertical lines.

```
0 0 0 0 0 0 0 0 0 0 0 0 0 0 0 0 0 0 0 0 0 0 0 0 0 0 017 0 0 0 0
0 0 0 0 0 0 0 0 0 0 0 0 0 0 0 0 0 0 0 0 0 0 0 0 0 0 0 0 0 0 0 0
0 0 0 0 0 0 0 0 0 0 0 0 0 0 0 0 0 0 0 0 0 0 0 0 0 0 0 0 0 0 0 0
0 0 0 0 0 0 0 0 0 0 0 0 0 0 0 0 0 0 0 0 0 0 0 0 0 0 0 0 0 0 0 0
0 0 0 0 0 0 0 0 0 0 0 0 0 0 0 0 0 0 0 0 0 0 0 0 0 0 0 0 0 0 0 0
0 0 0 0 0 0 0 0 0 0 0 0 0 0 0 0 0 0 0 0 0 0 0 0 0 0 0 0 0 0 0 0
0 0 0 0 0 0 0 0 0 0 0 0 0 0 0 0 0 0 0 0 0 0 0 0 0 0 0 0 0 0 0 0
0 0 0 0 0 0 0 0 0 0 0 0 0 0 0 0 0 0 0 0 0 0 0 0 0 0 0 0 0 0 0 0
0 0 0 0 0 0 0 0 0 0 0 0 0 0 0 0 0 0 0 0 0 0 0 0 0 0 0 0 0 0 0 0
0 0 0 0 0 0 0 0 0 0 0 0 0 0 0 0 0 0 0 0 0 0 0 0 0 0 0 0 0 0 0 0
0 0 0 0 0 0 0 0 0 0 0 0 0 0 0 0 0 0 0 0 0 0 0 0 0 0 0 0 0 0 0 0
0 0 0 0 0 0 0 0 0 0 0 0 0 0 0 0 0 0 0 0 0 0 0 0 0 0 0 0 0 0 0 0
0 0 0 0 0 0 0 0 0 0 0 0 0 0 0 0 0 0 0 0 0 0 0 0 0 0 0 0 0 0 0 0
0 0 0 0 0 0 017 0 0 0 0 0 0 018 0 0 0 0 0 0 017 0 0 0 0 0 0 016
0 0 0 0 0 0 0 0 0 0 0 0 0 0 0 0 0 0 0 0 0 0 0 0 0 0 0 0 0 0 0 0
0 0 016 0 0 0 0 0 0 018 0 0 0 0 0 0 018 0 0 0 0 0 0 016 0 0 0 0
0 0 0 0 0 0 0 0 0 0 0 0 0 0 0 0 0 0 0 0 0 0 0 0 0 0 0 0 0 0 0 0
0 0 0 0 0 0 017 0 0 0 0 0 0 019 0 0 0 0 0 0 017 0 0 0 0 0 0 015
0 015 0 016 0 0 0 018 0 019 0 0 0 019 0 018 0 0 016 0 0 015 0 0
014 015 0 01617 018 019 0 02020 020 019 018 017 016 015 014 014
0 014 015 0 0 0 0 019 020 0 0 0 0 020 019 0 0 0 0 015 014 0 0 0
0131314 0151617 0181920 021212121 02120 01918171615 014 0131313
1212 013141516171819 020212222222222221 02019181716151 4 013121212
1111121313141517181920212222232323232222211918171615141312111111
1010111213141517182021222324242424242423222120191716141312111110 10
9 9101112131517192021232425262626252523222119171614131110 9 9 9
7 7 81011131517182123242526272828272625232220181614121 0 9 8 7 7
6 5 6 8101214171922242527282930302827262422201815121 0 9 7 6 5 5
2 4 5 6 911141720212427293132323232292826232118141 1 9 7 5 4 3 3
1 2 2 4 71112 0 02327283132 32 0 0323231282522181512 8 5 3 2 1 1
0 0 1 2 5 0 0 0 0 0 03232 0 0 0 0 0 03229 0 0 0 0 0 5 3 1 0 0 0
0 0 0 0 0 0 0 0 0 0 0 0 0 0 0 0 0 0 0 0 0 0 0 0 0 0 0 1 0 0 0 0
```

FIGURE 6.6. POLCAR transform of vertical lines.

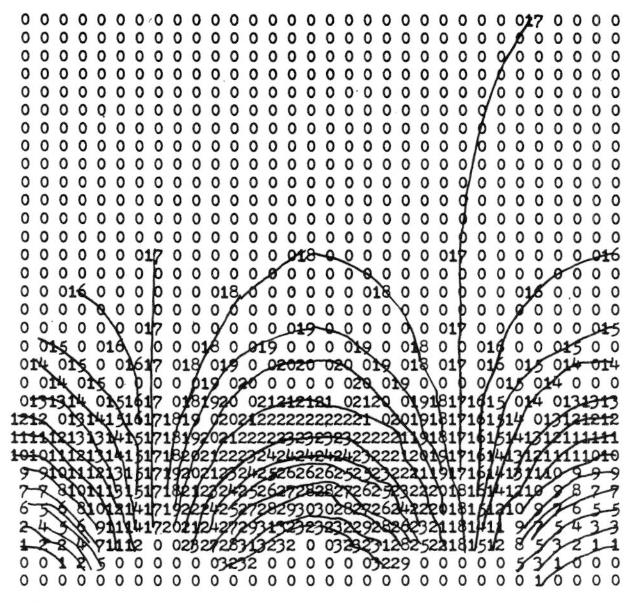

FIGURE 6.7. Isolines of POLCAR transform of vertical lines.

```
323131313131313030303030303030303030303030303030303031313131313131
313131313131303030303030303029292929292929292930303030303030313131
313131313130303030303029292929292929292929292929293030303030313131
313131303030303030292929292929292828282829292929292930303030303131
313130303030302929292929282828282828282828282829292929293030303031
313030303029292929282828282828272727282828282829292929303030303030
313030302929292928282828272727272727272727282828282929292930303030
303030302929292828282727272726262626262727272728282829292930303030
303030292929282828272727262626262626262627272728282829292930303030
303029292928282827272626262525252525252526262627272828282929293030
303029292928282727262626252524242424242425252626262727282829292930
303029292828282727262625242423232323232424252626272728282829292930
302929292828272726262524242322222222222232424252626272728282929292
302929292828272726252524232221202020212222324252526272728282929292
302929282828272626252423222120191819202122232425262627282828292929
302929282827272626252423222019161416192022232425262627272828292929
302929282827272626252423222018 14  1141820222324252626272728282929
302929282827272626252423222019161416192022232425262627272828292929
302929282828272626252423222120191819202122232425262627282828292929
302929292828272726252524232221202020212222324252526272728282929292
302929292828272726262524242322222222222232424252626272728282929292
303029292828282727262625242423232323232424252626272728282829292930
303029292928282727262626252524242424242425252626262727282829292930
303029292928282827272626262525252525252526262627272828282929293030
303030292929282828272727262626262626262627272728282829292930303030
303030302929292828282727272726262626262727272728282829292930303030
313030302929292928282828272727272727272727282828282929292930303030
313030303029292929282828282828272727282828282829292929303030303030
313130303030302929292929282828282828282828282829292929293030303031
313131303030303030292929292929292828282829292929292930303030303131
313131313130303030303029292929292929292929292929293030303030313131
313131313131303030303030303029292929292929292930303030303030313131
```

FIGURE 6.8. *IR* values of POLCAR transform as a function of *I*,*J*.

```
3231313131313130303030303030303030303030303030303030313131313131
3131313131303030303030302929292929292929293030303030303031313131
3131313131303030302929292929292929292929292929293030303031313131
3131313130303030292929292929282828282829292929292930303030313131
3131313030302929292929292828282828282828282929292929293030303131
3130303030292929292828282828282727272828282828282929292930303030
3130303029292929282828282727272727272727272828282829292929303030
3030303029292928282827272727262626262627272727282828292929303030
3030302929292828282727272626262626262626262727272828282929293030
3030292929282828272726262625252525252525262626272728282829292930
3030292929282827272626262525242424242425252626262727282829292930
3030292928282827272626252424232323232324242526262727282828292930
3029292928282727262625242423222222222223242425262627272828292929
3029292928282727262525242322212020202122232425252627272828292929
3029292828282726262524232221201918192021222324252626272828282929
3029292828272726262524232220191614161920222324252626272728282929
30292928282727262625242322201814 1141820222324252626272728282929
3029292828272726262524232220191614161920222324252626272728282929
3029292828282726262524232221201918192021222324252626272828282929
3029292928282727262525242322212020202122232425252627272828292929
3029292928282727262625242423222222222223242425262627272828292929
3030292928282827272626252424232323232324242526262727282828292930
3030292929282827272626262525242424242425252626262727282829292930
3030292929282828272726262625252525252525262626272728282829292930
3030302929292828282727272626262626262626262727272828282929293030
3030303029292928282827272727262626262627272727282828292929303030
3130303029292929282828282727272727272727272828282829292929303030
3130303030292929292828282828282727272828282828282929292930303030
3131313030302929292929292828282828282828282929292929293030303131
3131313130303030292929292929282828282829292929292930303030313131
3131313131303030302929292929292929292929292929293030303031313131
3131313131303030303030302929292929292929293030303030303031313131
```

FIGURE 6.9. Isolines of IR values.

```
2828282727272727262626262525252424232323232222222222212121212020
2828282827272727272626262525252424232323222222222221212121202020
2828282828272727272626262525252424232323222222222121212121202020
2928282828282727272726262625252424242323232222222221212120202020
2929282828282827272726262625252424242323222222222121212020202020
2929292828282827272727272626252524242323222222222121212020202019
2929292928282828272727262626252524242323222222212120202020191919
2929292929282828282727272626252524242322222222212020202019191919
3029292929292928282827272626252524232322222221202020201919191919
3030302929292929282828272726252524232322212120202019191919191918
3030303030302929292928282827262524232222212120201919191919181818
3030303030303029292928282727262524232221212020191919191818181818
3131313030303030302929292827262524232221202019191918181818181818
3131313131313030303030292928272624222120191919181818181817171717
3131313131313131313131303030292826242220191918181817171717171717
3232323231313131313131313131303028242019181717171717171717171717
3232323232323232323232323232322228161616161616161616161616161616
 1 1 1 1 1 1 1 1 1 1 1 1 2 2 3 4 8121415151515161616161616161616
 1 1 1 1 1 1 1 2 2 2 2 2 3 3 4 6 8111213141415151515151515151616
 1 1 2 2 2 2 2 2 2 3 3 3 4 4 5 7 8101112131414141415151515151515
 2 2 2 2 2 2 2 3 3 3 3 4 4 5 6 7 8101112121313141414141415151515
 2 2 2 2 2 3 3 3 3 4 4 4 5 6 6 7 8 91011121213131314141414141415
 2 2 3 3 3 3 3 3 4 4 4 5 5 6 7 8 8 91011111212131313141414141414
 3 3 3 3 3 3 4 4 4 4 5 5 6 6 7 8 8 91010111112121313131314141414
 3 3 3 3 3 4 4 4 4 5 5 6 6 7 7 8 8 91010111112121213131313131414
 3 3 3 4 4 4 4 4 5 5 5 6 6 7 7 8 8 9 910101111121212131313131314
 3 3 4 4 4 4 4 5 5 5 6 6 6 7 7 8 8 9 910101111111112121213131313
 4 4 4 4 4 4 5 5 5 6 6 6 7 7 7 8 8 9 910101011111112121212131313
 4 4 4 4 4 5 5 5 5 6 6 6 7 7 8 8 8 9 910101011111112121212121313
 4 4 4 4 5 5 5 5 6 6 6 7 7 7 8 8 8 9 9 9101011111111111212121213
 4 4 4 5 5 5 5 5 6 6 6 7 7 7 8 8 8 9 9 9101010111111111112121212
 4 4 5 5 5 5 5 6 6 6 6 7 7 7 8 8 8 9 9 9101010111111111112121212
```

FIGURE 6.10. Ith values of POLCAR transform as a function of I,J.

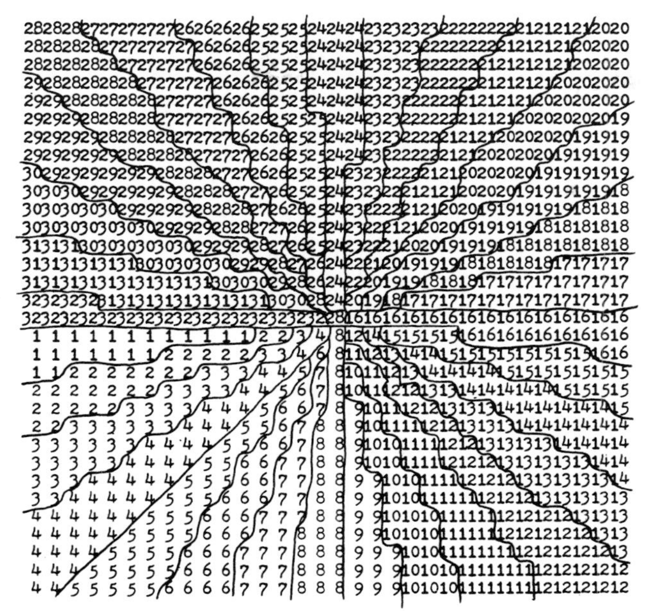

FIGURE 6.11. Isolines of *ITH* values.

The form of *IR* is

$$IR = H \log R + K$$

H and *K* are chosen so that the minimum value of *R* is 1 and the maximum value of *IR* is 32. The logarithm is employed to provide the scale insensitivity of the Mellin transform. The gray-level values are amplified by a factor of R^2 since in the frequency domain this is comparable to taking a second derivative of the spatial domain input image to emphasize object edges.

Shape Detection Algorithm

A block diagram of a shape detection algorithm incorporating the POLCAR transform is shown in Table 6.2. Two images, *I*1 and *I*2, are input. Their power spectra are obtained using a FFT subroutine (see Appendix). The output FFT has the zero-frequency component at (1,1); to simulate an optical power spectra, a rearrangement of the frequency components with the zero-frequency component at (17,17) is effected. The output, aside from various

TABLE 6.1
POLCAR Subroutine

```
  SUBROUTINE POLCAR (IMAG,N)
  DIMENSION IMAG(32,32),IA(32,32)
  XX=FLOAT(N)/2.+1.01
  DO 1 I=1,N
  DO 1 K=1,N
1 IA(I,K)=0
  DO 8 I=1,N
  Y=-FLOAT(I)+XX
  DO 8 J=1,N
  X=FLOAT(J)-XX
  R=X*X+Y*Y
  R=SQRT(R)
  R=R+.1
  IR=ALOG10(R)*13.155+14.2
  TH=ATAN2(Y,X)+3.14159
  ITH=5.*TH+1.
  IF (IR.LT.1.OR.ITH.LT.1) GO TO 8
  KR=R-.1
  IMAG(I,J)=IMAG(I,J)*KR*KR
  IA(IR,ITH)=IMAG(I,J)
8 CONTINUE
  DO 7 I=1,N
  DO 7 J=1,N
7 IMAG(I,J)=IA(I,J)
  RETURN
  END
```

intermediate displays, is a value of QMAX ($I1,I2$). This is used to obtain $M(I1,I2)$.

Discussion

Sixteen binary 32 × 32 arrays containing simple objects were studied. There are essentially six types of objects categorized by shape in the set of test images:

1. Solid rectangles, 2.6:1 aspect ratio; $I1$, $I2$, $I5$, $I6$.
2. Solid right isosceles triangles; $I3$, $I4$, $I10$.
3. Solid rectangles, 3.0:1 aspect ratio; $I7$, $I8$, $I9$.
4. Solid circles; $I11$, $I12$

TABLE 6.2
Flow Diagram of Shape Detection Algorithm

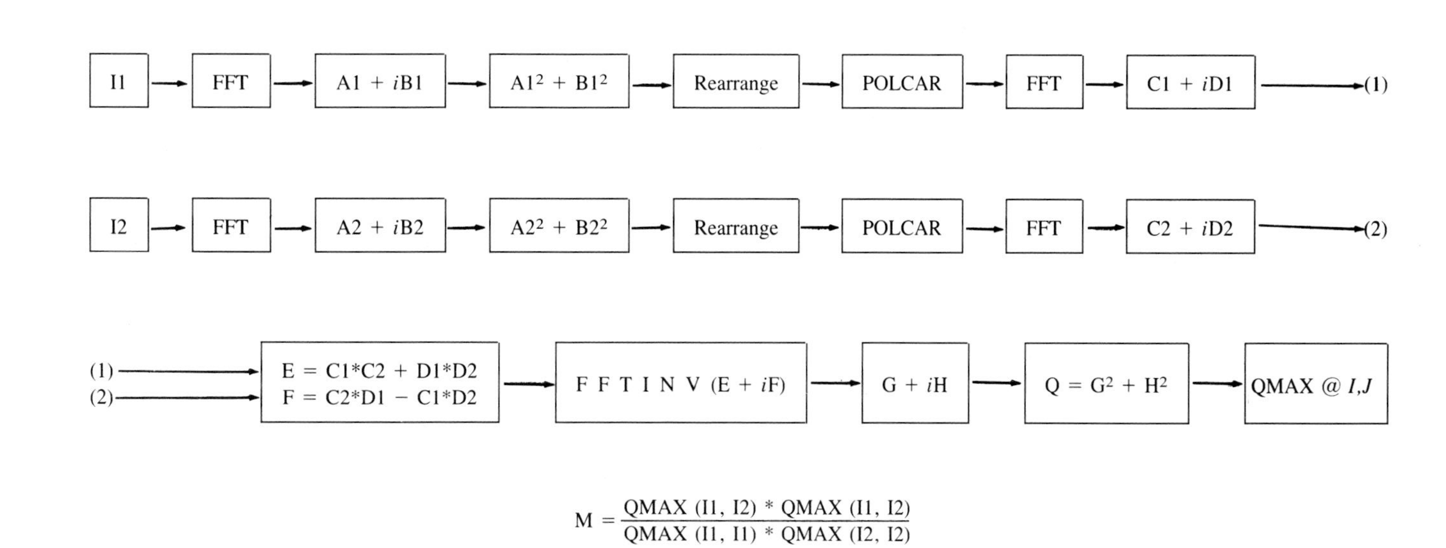

$$M = \frac{\text{QMAX (I1, I2)} * \text{QMAX (I1, I2)}}{\text{QMAX (I1, I1)} * \text{QMAX (I2, I2)}}$$

TABLE 6.3
M values for 32 × 32 shape comparisons

	*I*1	*I*2	*I*3	*I*4	*I*5	*I*6	*I*7	*I*8	*I*9	*I*10	*I*11	*I*12	*I*13	*I*14	*I*15	*I*16
I1	1.000															
I2	.103	1.000														
I3	.018	.078	1.000													
I4	.020	.057	.797	1.000												
I5	.068	.344	.087	.071	1.000											
I6	.102	.999	.078	.057	.346	1.000										
I7	.131	.925	.071	.057	.317	.923	1.000									
I8	.310	.177	.048	.039	.122	.176	.198	1.000								
I9	.154	.237	.029	.048	.089	.237	.278	.080	1.000							
I10	.026	.043	.461	.347	.056	.044	.037	.061	.021	1.000						
I11	.018	.078	.091	.108	.035		.060	.031	.053	.078	1.000					
I12	.010	.040	.132	.136	.041		.035	.015	.028	.137	.276	1.000				
I13	.032	.056	.327	.328	.048		.055	.076	.176	.100	.053	.086	1.000			
I14	.036	.024	.077	.083	.020		.027	.033	.041	.239	.047	.087	.264	1.000		
I15	.215	.307	.021	.023	.045		.350	.270	.253	.018	.033	.029	.101	.066	1.000	
I16	.087	.116	.029	.042	.207		.130	.143	.109	.012	.015	.013	.136	.036	.233	1.000

5. Hollow right isosceles triangles; *I*13, *I*14
6. Hollow rectangles, 3.0:1 aspect ratio; *I*15, *I*16

The values of *M* obtained by comparing various image pairs using the program called SHAPE, are shown in Table 6.3. High values indicate a good match, low values a poor match. *I*6 is simply *I*2 with gray values multiplied by 4; the *M* values in Table 6.3 for *I*6, as expected, are virtually identical to those for *I*2. Hence *I*6 need not be analyzed as a distinct object for these purposes.

For the 15 distinct objects, 105 cross-comparisons are possible. It is expected that objects with shapes of the same type should have high *M* values. Naturally, all objects have somewhat the same shape as all other objects. Hence *M* can take on a continuous set of values from zero to 1. For discrimination purposes, a threshold, *T*, can be specified; here $T = .080$ is chosen, that is, *M* is considered high if it is greater than or equal to *T*.

Tables 6.4 through 6.9 exhibit subsets of Table 6.3 for the six object types. Of the 12 comparisons shown, 11 detections succeed, but 1 falls below the threshold, *T*, that for *I*1 versus *I*5. This comparison involves a scale factor of 2, together with a 90° rotation and a translated centroid.

There are, other than these expected high *M* values, 36 cross-object high *M* values. Most of these are not too disturbing. Twenty are rectangles matched to rectangles but solid versus hollow or of differing aspect ratio. Five are triangles matched to triangles, solid versus hollow. Seven spurious high *M*

TABLE 6.4
Object A

	*I*1	*I*2	*I*5
*I*1	1.0	—	—
*I*2	.103	1.0	—
*I*5	.068	.344	1.0

TABLE 6.5
Object B

	*I*3	*I*4	*I*10
*I*3	1.0	—	—
*I*4	.797	1.0	—
*I*10	.461	.347	1.0

TABLE 6.6
Object C

	*I*7	*I*8	*I*9
*I*7	1.0	—	—
*I*8	.198	1.0	—
*I*9	.278	.080	1.0

TABLE 6.7
Object D

	*I*11	*I*12
*I*11	1.0	—
*I*12	.276	1.0

TABLE 6.8
Object E

	*I*13	*I*14
*I*13	1.0	—
*I*14	.264	1.0

TABLE 6.9
Object F

	*I*15	*I*16
*I*15	1.0	—
*I*16	.233	1.0

values match triangles with circles; of these seven, five are solid triangles, two are hollow triangles. Discrimination is shown to be unreliable in these 32 × 32 format cases. The four remaining anomalous high M values erroneously match triangles and rectangles. The other 47 M values of Table 6.3 are all less than T.

Format Size (64 × 64)

The digital image shape detection research for binary images of size 32 × 32, has been extended to address image arrays measuring 64 × 64. The

FORTRAN SHAPE program described previously was used with minor modification in this 64 × 64 study. These modifications are

1. Substitute 64 for every occurrence of 32 in the main program and in all subroutines expect the FFT.
2. In the POLCAR subroutine, change to

$$IR = AL0G10(R) * 24.0 + 24.1$$
$$ITH = 10. * TH + 1$$

3. Delete array output displays to reduce wall time.

Sixteen binary 64 × 64 arrays containing simple objects were studied. The values of M obtained by comparing various (64 × 64) image pairs using the SHAPE program are shown in Table 6.10. There are essentially five types of objects categorized by shape in the set of test images:

1. Solid rectangles, 3.0:1 aspect ratio: $I3$, $I4$, $I5$, $I7$, $I10$, $I12$
2. Solid right isosceles triangles: $I1$, $I2$, $I9$, $I11$
3. Solid circles: $I6$, $I8$, $I15$
4. Unequal arm crosses: $I13$, $I14$
5. Random noise: $I16$

For the distinct objects, 120 cross-comparisons are possible. It is noted that, in contrast to the 32 × 32 study, the average M value is lower by one-half. The mean of the 105-entry $M32$ table is .122; the mean of the 120-entry $M64$ table is .061. Accordingly, a lower threshold for discrimination purposes can be selected; here $T = .056$ is selected, whereas $T = .080$ was used in the 32 × 32 case.

Tables 6.11–6.14 exhibit subsets of Table 6.10 for four of the five object types. No table is needed for object type E since, as expected, none of the other objects produce a high M value when compared with $I16$, random noise.

Of the 25 comparisons shown, 22 succeed, but 3 fall below the threshold, T. The three failures are all for $I5$. $I5$ is a vertical rectangle, and matches are found for the other vertical rectangles, $I7$ and $I12$, but not for the horizontal rectangles, $I3$ and $I4$, nor for the 45° rectangle, $I10$. The other vertical rectangles do find matches with the horizontal rectangles and the 45° rectangle.

There are, other than the 25 expected high M values, 11 cross-object high M comparisons. Of these 11 false alarms, 7 are for object type D. $I13$ is mistaken for vertical rectangles three times and once for a triangle; $I14$ is confused twice for horizontal rectangles and once for the 45° rectangle. Clearly, this algorithm is not suited for discriminating unequal arm crosses from other simple objects. The other four false alarms are triangle–rectangle mismatches.

TABLE 6.10
M values for 64 × 64 shape comparisons

	*I*1	*I*2	*I*3	*I*4	*I*5	*I*6	*I*7	*I*8	*I*9	*I*10	*I*11	*I*12	*I*13	*I*14	*I*15	*I*16
I1	1.000															
I2	.299	1.000														
I3	.025	.045	1.000													
I4	.021	.064	.201	1.000												
I5	.032	.031	.032	.033	1.000											
I6	.050	.055	.008	.008	.011	1.000										
I7	.028	.025	.204	.059	.164	.010	1.000									
I8	.051	.051	.005	.005	.011	.076	.006	1.000								
I9	.205	.217	.190	.106	.028	.035	.037	.035	1.000							
I10	.039	.145	.106	.311	.022	.002	.057	.003	.391	1.000						
I11	.806	.162	.032	.027	.038	.045	.034	.046	.216	.038	1.000					
I12	.038	.018	.066	.231	.154	.009	.232	.006	.021	.150	.038	1.000				
I13	.054	.052	.030	.031	.123	.008	.125	.003	.015	.028	.060	.107	1.000			
I14	.007	.025	.081	.109	.025	.003	.045	.001	.036	.174	.010	.050	.113	1.000		
I15	.038	.041	.003	.002	.005	.169	.004	.103	.028	.004	.033	.002	.003	.002	1.000	
I16	.006	.005	.004	.003	.002	.003	.003	.007	.003	.004	.004	.002	.027	.015	.005	1.000

TABLE 6.11
Object A

	*I*3	*I*4	*I*5	*I*7	*I*10	*I*12
*I*3	1.000	—	—	—	—	—
*I*4	.201	1.000	—	—	—	—
*I*5	.032	.033	1.000	—	—	—
*I*7	.204	.059	.164	1.000	—	—
*I*10	.106	.311	.022	.057	1.000	—
*I*12	.066	.231	.154	.232	.150	1.000

TABLE 6.12
Object B

	*I*1	*I*2	*I*9	*I*11
*I*1	1.000	—	—	—
*I*2	.229	1.000	—	—
*I*9	.205	.217	1.000	—
*I*11	.806	.162	.216	1.000

TABLE 6.13
Object C

	*I*6	*I*8	*I*15
*I*6	1.000	—2	—
*I*8	.076	1.000	—
*I*15	.169	.103	1.000

TABLE 6.14
Object D

	*I*13	*I*14
*I*13	1.000	—
*I*14	.113	1.000

It is satisfying to note that the circle–triangle discrimination problem encountered in the 32×32 case has, in the 64×64 case, gone away. All circles are detected, no circles are false alarmed. The other 84 M values of Table 6.10 are all less than T. No attempt at minimizing execution time was made in programming. Execution time for the 32×32 SHAPE program on a PDP–10 computer operating under the TENEX operating system is approximately 17 sec; the 64×64 SHAPE program executes in approximately 23 sec.

ENHANCING DIGITAL IMAGES FOR MAXIMUM INTERPRETABILITY USING LINEAR PROGRAMMING[2]

Statement of the Problem

It has for some time been recognized that digital image enhancement offers considerable payoff in terms of photointerpretation utility. Many degrading effects can be compensated, and the information extraction process can be assisted in no small measure. Commonly, for example, motion-induced image smear, defocus, poor contrast, and geometric distortion are treated.

Many enhancement techniques have been reported,[3] often employing Fourier analysis (Gezari, Labeyrie, & Stachnik, 1972; Harris, 1966; Knight, 1973; McGlamery, 1967; Papoulis, 1972). In general, some effort is devoted to determining the point-spread function (PSF) using a variety of optical procedures. At some point, the current best estimate of the PSF is hypothecated as suitable and is used in the deconvolution process. The image array **I** is assumed to be given by

$$\mathbf{I} = \mathbf{S}*\mathbf{O} + \mathbf{N} \tag{8}$$

where **S** is the PSF (taken as known), **N** is noise (unknown), **O** is the object radiance field that generated the image (sought), and the asterisk indicates convolution.

In some cases, the noise term is neglected. The procedure is to solve Eq.(8) by dividing the Fourier transform of the image array by the Fourier transform

[2]This section previously appeared in Hord, *Computer Graphics and Image Processing*, 1977, *6*, 295–306.

[3]Since this work was first reported in Itek's IR&D report, related studies have appeared (e.g., Mascarenhas, 1974).

of the PSF and obtaining the inverse Fourier transform of the ratio. Other deconvolution procedures have been used, including relaxation, eigenfunctions (Barnes, 1966), and variational (Frieden, 1972; Frieden & Burke, 1972) (Lagrange multiplier) algorithms. Aside from deconvolution, other enhancement algorithms (Bayes' theorem [Helstrom, 1967; Richardson, 1972]), decision theory [Harris, 1964*a*] analytic extension [Harris, 1964*b*], and pattern recognition schemes [Montanari, 1971]) have been reported. Although these are not reviewed here, some shortcomings of these methods can be specified. These often include

1. No quantitative measure of success
2. Inaccuracies associated with PSF neglected
3. Noise often neglected
4. Nonmeaningful object features enhanced
5. Multiple isoplanatic patches treated as one
6. No systematic way to incorporate a priori knowledge
7. Multiple images of same target enhanced independently

Approach

We will now describe an approach that is not subject to the shortcomings just listed. The interpreter does not necessarily want an exact reconstruction of the original object radiance field, particularly if that incorporates glint, camouflage, or some feature that reduces interpretability. The desired result, instead, is some transformed image that exhibits maximum interpretability of the content for exploitation. Although a perfect quantitative measure of image interpretability has not yet been found, many image quality measures are highly correlated with photointerpreter performance. These include: (*a*) minimum noise, (*b*) edge gradient, (*c*) relative structural content, (*d*) fidelity defect, (*e*) correlation quality, (*f*) sharpness, (*g*) acutance, (*h*) contrast, (*i*) entropy, (*j*) maximum likelihood, and (*k*) mean square error.

Each of these image quality measures is defined in terms of a mathematical expression and can be evaluated objectively and quantitatively. It must be noted that because no quality measure is perfectly correlated with interpreter performance, the interpreter should be provided with several renderings of the image, when possible. Each may exhibit a quality not available in the other versions of the picture. For example, in two images of the scene, objects may more easily be identified in one, whereas the other exhibits better resolution for detail.

1	2
3	4

FIGURE 6.12. Object distribution (1).

The approach, then, is to enhance images, not with the image-generating object as a goal, but with maximized image quality measures as the goal, each measure producing, perhaps, a different version of the image. This set of enhancements is expected to have improved interpretability over the raw image and over images enhanced by other schemes. Here linear programming has been considered as the mathematical algorithm for the enhancement process. The image elements constitute the right-hand side of the constraint equations, the object elements are the variables, and the PSF elements correspond to the coefficients. The noise is absorbed in the slack variables. A 2×2 object, for example, convolved with a 3×3 PSF producing a 4×4 image gives rise to 16 constraint equations.

If linear programming can be shown to be effective in the enhancement of digital imagery, then it should be noted that certain features of linear programming make it attractive. Some of these potential benefits are as follows:

1. The figure of merit, related to interpretability, is optimized.
2. A priori knowledge is systematically incorporated as additional constraint equations.
3. It is iterative in nature, with each iteration improving on prior iterations.
4. Sensitivity analysis is available.
5. A convenient change of figure of merit and/or constraints is possible.
6. A positive definite nature of PSF and object is inherent in this method.
7. It has a natural role for additive noise as slack variables.
8. Sophisticated software is currently available.
9. It makes it possible to solve for PSF under constraints.

2	2
10	10

FIGURE 6.13. Object distribution (2).

4	4	4	4	4	4	4	4	4	4
4	4	4	4	4	4	4	4	4	4
4	4	4	4	4	4	4	4	4	4
4	4	4	4	4	4	4	4	4	4
4	4	4	4	4	4	4	4	4	4
8	8	8	8	8	40	40	40	40	40
8	8	8	8	8	40	40	40	40	40
8	8	8	8	8	40	40	40	40	40
8	8	8	8	8	40	40	40	40	40
8	8	8	8	8	40	40	40	40	40

FIGURE 6.14. Object distribution (3).

10. It provides superresolution Frieden (1972).
11. Simultaneous processing of multiple images is possible.

Experiments

INPUT

The linear programming model has been tested with three test object distributions designated (1), (2), and (3). Distribution (1) is shown in Figure 6.12. The contrasts between adjacent elements vary from 3:1 to 1.3:1. The second object (Figure 6.13) is a horizontal edge with a 5:1 contrast. The final object distribution (Figure 6.14) consists of a 10 × 10 array exhibiting three well-defined edges with 10:1, 5:1, and 2:1 contrasts. The 2 × 2 objects were convolved with the 3 × 3 PSF shown in Figure 6.15*a*, whereas the PSF convolved with object (3) is shown in Figure 6.15*b*. Convolving the 2 × 2 objects produces 4 × 4 arrays, whereas the 10 × 10 object generates a 12 × 12 array. Adding correspondingly dimensioned noise arrays to these convolutions results in the hypothesized images, the data on which the enhancement programs operate.

Zero Noise Cases

In evaluating a reconstruction method, a reasonable starting place is the examination of how well the method restores noiseless data. Deconvolving a noiseless image with the exact PSF that generated it should, if the algorithm is

(a)			(b)		
1	2	1	0.07513	0.12402	0.07513
2	3	2	0.12402	0.20340	0.12402
1	2	1	0.07513	0.12402	0.07513

FIGURE 6.15. (*a*) The 3 × 3 PSF. (*b*) The PSF convolved with object (3).

effective, recover the original object distribution. In all zero noise cases tested with all merit functions, the linear programming model reconstructed the original object distribution exactly.

Figures of Merit

Four merit functions were considered in the linear programming model study:

1. Equal weight to the object points and no penalty or cost for noise. (For example, $F = O_{11} + O_{12} + O_{21} + O_{22}$ for object [1].)
2. Equal weight to object points with noise penalized. (For example, $F = O_{11} + O_{12} + O_{21} + O_{22} - N_{11} - N_{21} - N_{22}$ for object [1].)
3. Weight the object points by a guess of their proper values with noise penalized. ($F = O_{11} + 2O_{12} + 3O_{21} + 4O_{22} - N_{11} - N_{12} - N_{21} - N_{22}$ for object [1].)
4. Equal weight on the object points and heavy noise penalty. ($F = O_{11} + O_{12} + O_{21} + O_{22} - 10[N_{11} + N_{12} + N_{21} + N_{22}]$.)

Since linear programming excludes nonlinear equations, considering nonlinear figures of merit was deferred. Signal-to-noise considerations motivated the chosen figure of merit designs. If one considers the output array O_{ij} to be an enhanced image rather than the object array, one would like as much of the available image energy as is consistent with the constraints to be assigned to the enhanced image. Obviously, if one penalized the Os and heavily weighted the Ns, the result would be all zeros. The chosen figures of merit are not considered optimal.

Fourier Comparison

On the 10×10 object distribution without noise, the linear programming model achieved a perfect reconstruction. This is a very important result because a Fourier reconstruction was unable to recover the object exactly.[4] The 10×10 object convolved with the 3×3 PSF was used as the input array. This array was Fourier transformed using the FFT. The PSF was transformed

[4]With perfect knowledge of the PSF, no zero PSF frequency components, and no noise, the FFT method will reconstruct the object exactly; however, perfect knowledge, in practice, is not available (Lerman, personal communication, 1972).

and the object spectrum was divided by this optical transfer function (OTF). The frequency array had the high frequencies boosted to increase the edge response in the best Fourier reconstruction. The reconstructed object array is shown in Figure 6.16.

This is not a perfect reconstruction because most of the edge information was not sampled at a fine enough increment to be preserved. The high frequencies are lost in the frequency domain, and even the relative enhancement of what high-frequency information we do have is not sufficient to provide a really effective reconstruction.

Perfect Reconstructions with Noise

Image 1 (object [1] convolved with the PSF, Figure 6.15a) was subjected to a 10% gain. This image distribution was recovered exactly with the linear programming model. It is possible for a large spike to be added to the input image, perhaps by a scratch, dust, or electronic malfunction. A large (199 gray level units) spike was added to the corner of the input image with no other noise and to the image, with a 10% gain. Again the reconstructions were exact. The second image had a diagonal of 1s added to it, and again the reconstruction was perfect.

1.	.29	.97	1.3	1.3	1.2	.85	-.11	-.52	-.52	-.53
2.	.73	2.5	3.2	3.2	3.2	2.9	2.3	2.1	2.1	2.0
3.	1.1	3.6	4.7	4.7	4.7	5.6	7.8	8.7	8.7	8.7
4.	.84	2.8	3.6	3.6	3.7	3.2	2.0	1.5	1.4	1.5
5.	.74	2.5	3.2	3.2	3.2	2.3	0.03	-.91	-.83	-.94
6.	1.4	4.9	6.3	6.4	6.4	9.9	18.	22.	22.	22.
7.	1.7	5.9	7.5	7.7	7.8	14.	29.	36.	36.	36.
8.	1.5	5.0	6.4	6.5	6.7	12.	26.	32.	32.	32.
9.	2.1	7.1	9.0	9.2	9.4	18.	37.	46.	46.	45.
10.	2.1	7.1	9.1	9.3	9.5	18.	37.	46.	46.	46.

FIGURE 6.16. Results of Fourier reconstruction (noise-free).

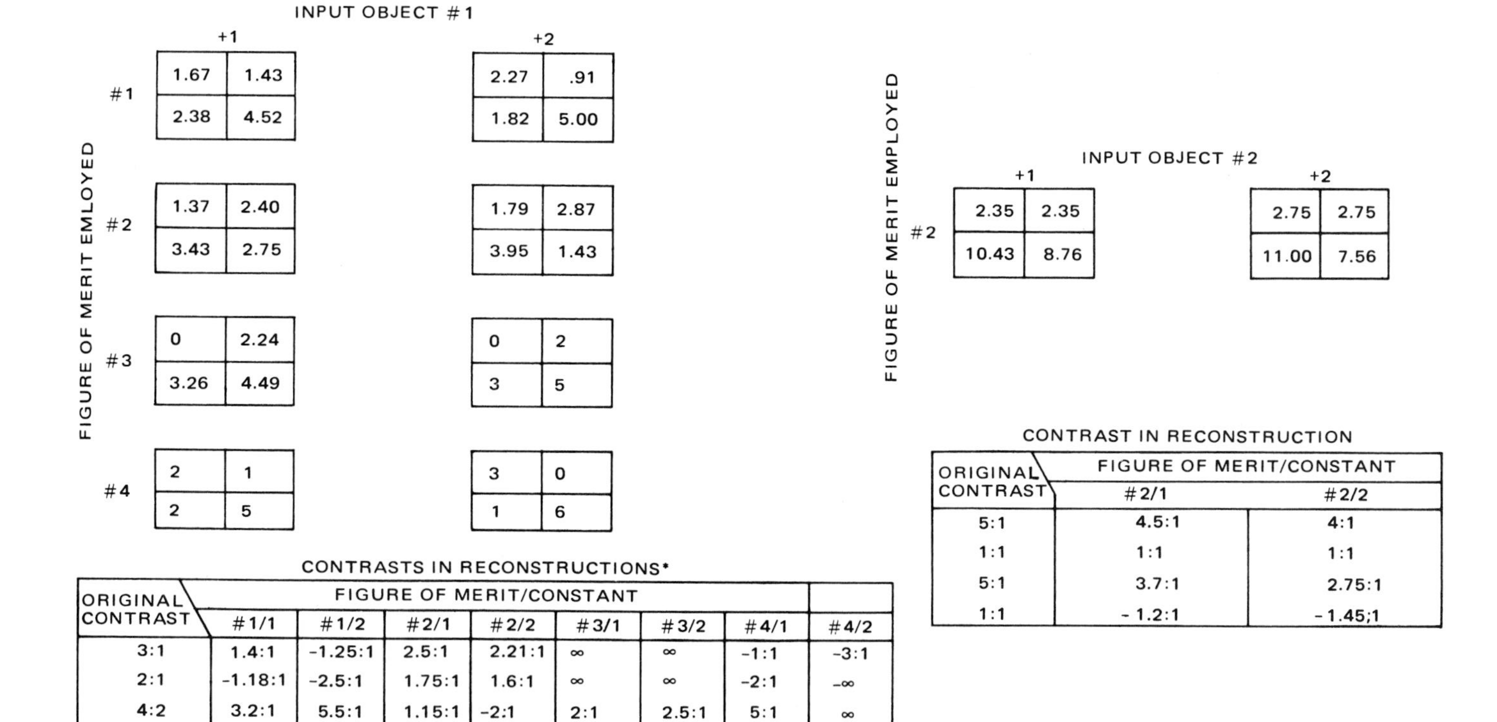

CONTRASTS IN RECONSTRUCTIONS*

ORIGINAL CONTRAST	FIGURE OF MERIT/CONSTANT							
	# 1/1	# 1/2	# 2/1	# 2/2	# 3/1	# 3/2	# 4/1	# 4/2
3:1	1.4:1	-1.25:1	2.5:1	2.21:1	∞	∞	-1:1	-3:1
2:1	-1.18:1	-2.5:1	1.75:1	1.6:1	∞	∞	-2:1	-∞
4:2	3.2:1	5.5:1	1.15:1	-2:1	2:1	2.5:1	5:1	∞
13:1	1.9:1	2.7:1	-1.25:1	-2.75:1	1.38:1	1.6:1	2.5:1	6:1

CONTRAST IN RECONSTRUCTION

ORIGINAL CONTRAST	FIGURE OF MERIT/CONSTANT	
	# 2/1	# 2/2
5:1	4.5:1	4:1
1:1	1:1	1:1
5:1	3.7:1	2.75:1
1:1	- 1.2:1	- 1.45;1

FIGURE 6.17. Reconstructions (input objects [1] and [2]) with additive constants. A negative contrast is one in the opposite direction from the original object.

Constant Noise

To simulate the effects of haze (essentially glare or fog since the operation was done on the perfect image after convolution with the PSF), a constant was added to each image point. There were two constants used: 1 and 2. This reduces the 3:1 edge to a 2:1 edge (with constant = 1) or a 1.67:1 edge (with constant = 2). The 1.3:1 contrast edge is reduced to either a 1.25:1 or a 1.2:1 contrast edge.

The results are summarized in Figure 6.17. The most important factors are the negative contrasts. This indicates contrasts in the reconstructed image that are in the reverse sense from those in the original image. Now the interpretation becomes more difficult. Looking at figure of merit (2) with an additive constant of 2, the fourth point of object (1) is definitely bad. This caused two contrasts to be reversed. With the same figure of merit but with an additive constant of 1, the fourth point is not quite as bad and reverses the sense of only one contrast edge. On the other hand, in the case of figure of merit (3), all the contrasts are in the correct sense. The reconstructed object looks very much like the original. Figure 6.18 shows the RMS error for the first object. In this case the root mean square (RMS) error is a useful quality criterion.

The second object was reconstructed with figure of merit (2), and the results were much more promising. There are two points of interest. The first is that the reconstruction is nonsymmetric. This is an artifact of the linear programming model used. The program works from the upper-left corner of the array to the lower-right corner of the image. It is very likely that if the image were rotated 180°, again reconstructed and the two results averaged, the nonsymmetric bias would be reduced. The second point is that the second object has a smaller percentage of error than the first. This is an expected result since the first object has four low-contrast edges whereas the second object consists of a single high-contrast edge. The higher signal:noise ratio of the second object accounts for the better reconstruction. A possible technique for handling the additive constant noise might be an iterative technique that maximizes the quality function by subtracting or adding a constant from the image array.

	Figure of merit.			
Added Constant	#1	#2	#3	#4
1	.59	.71	.34	2.00
2	1.14	1.49	.71	2.00

FIGURE 6.18. RMS error (object [1]).

Constant Noise with Spike

Object (1) had both an additive constant and a large spike added to the image, and figure of merit (4) was used. The results are shown in Figure 6.19. It can be seen that the linear programming solution can handle this case accurately and even improves the reconstruction over the additive constant without the spike. It should be noted that the occurrence of large noise spikes in the input image array is not a problem that is handled well by Fourier techniques since there is a conservation of energy in the Fourier technique and a large spike tends to be spread throughout the image, reducing the contrasts, etc.

Random Noise

A much more practical noise situation is one in which random noise is added to the input image. An important test of any reconstruction scheme is its ability to handle random noise. Object (2) had Gaussian random noise systematically added to the input image. The noise did not have a zero mean since the linear programming model requires positive, definite variables. For a thorough test, therefore, noise with the same statistics (identically distributed) was added four separate times. There are two different noise cases ($\bar{X}$ of .1, σ of .1, and $\bar{X}$ of .2, $\sigma = .1$). These eight cases were processed with figure of merit (3). The reconstruction was excellent. The average RMS error (normalized to the input object power) was 4.5×10^{-3} for the first noise type and 7.5×10^{-3} for the second noise type. The first noise type was added to object (1) and the average RMS error was 7×10^{-2}, almost four times as

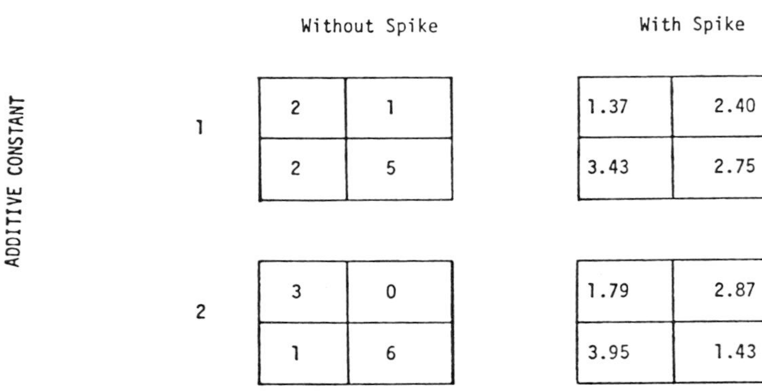

FIGURE 6.19. Figure of merit (4); object (1).

great as for object (2). This is likely due to the low signal content of object [1].

To provide a more realistic evaluation of the linear programming model, a larger (10 × 10) object, [3], was used. This object, because of its large size, was only used for a limited number of trials. To this input image was added Gaussian random noise of two statistical types: $\bar{X}$ of 1.0, σ of .5 and $\bar{X}$ of .2, σ of .05.) This was done twice for each noise type. A reconstructed object distribution is shown in Figure 6.20 for the second, (.2, .05) noise type.

To determine the effectiveness of the reconstruction model, each of the three reconstructed edge distributions was compared against the original edge distribution. Since each edge has five traces across it, these traces were averaged. Figure 6.21 shows a plot of the averaged traces across the 2:1 edge in the (.2, .05) noise reconstruction. In the σ of .05 case, the noise is smoothed out for all contrasts and the edge is easily observable. For the noise σ of .5, all but the 2:1 contrast edge are easily observable and even that edge is visible though not obvious.

Fidelity Defect Quality Measure

A fidelity defect was defined for the edges as:

$$F = \left[\sum_i \frac{(x_i - o_i)^2}{o_i} \right]^{\frac{1}{2}}$$

	1	2	3	4	5	6	7	8	9	10
1	6.71	1.94	7.19	3.09	5.03	6.02	2.53	7.57	2.09	6.68
2	2.52	4.18	3.59	3.69	2.83	4.03	3.67	1.72	5.09	2.05
3	5.97	4.92	4.68	4.25	7.58	.96	8.01	4.87	2.79	6.81
4	4.38	.14	7.72	0.0	4.76	6.31	0.0	5.42	5.10	3.03
5	3.93	6.71	1.21	9.24	0.0	5.71	3.72	5.81	.58	5.88
6	11.56	4.98	10.62	6.49	8.82	41.22	39.70	39.59	41.46	41.07
7	4.56	8.01	10.76	2.45	13.02	36.98	39.67	42.33	38.58	38.69
8	13.46	7.80	4.86	16.79	0.0	45.35	39.14	40.75	38.77	43.83
9	4.82	7.24	11.66	.24	15.07	36.38	40.15	40.77	40.25	37.85
10	10.74	6.42	9.75	9.22	7.27	41.51	39.46	42.10	37.74	43.19

FIGURE 6.20. Reconstruction with noise.

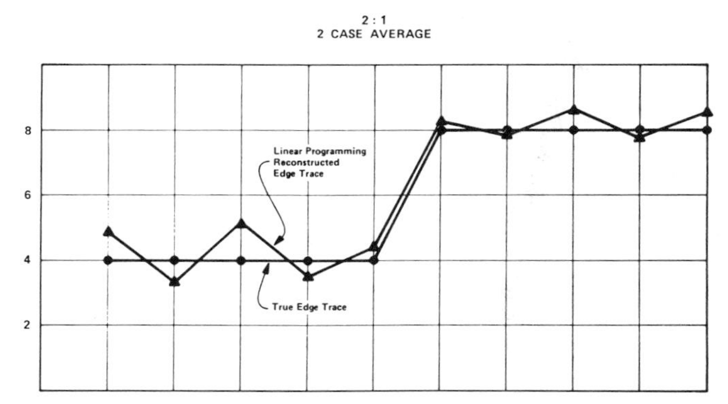

FIGURE 6.21. Noise: $\bar{X} = .2$, $\sigma = .05$.

Figure 6.22 shows the fidelity defect values for the observed cases. As the noise increases, the fidelity defect increases, as expected. Using the fidelity defect measure, the linear programming reconstruction model can be compared with the noise-free Fourier reconstruction method. The line edges have been smoothed by averaging in the same manner as before, and the 2:1 edge is shown in Figure 6.23. The fidelity defect is shown along with the linear programming values in Figure 6.22. It can be easily seen that the Fourier reconstruction is not as successful as the linear programming model.

Discussion

This section has tried to assess the viability of a series of contentions that together constitute a new approach to image enhancement. Feasibility has been demonstrated, and, to some extent, superiority over classical techniques has been shown as well. Of particular note are the perfect restorations achieved with no noise, a spike, and a diagonal line superimposed. This suggests that this scheme is suited to recontruct images that have been acquired or scanned electronically, since these kinds of effects are often due to electrical or mechanical malfunctions. Note, also, that the reconstruction is independent of the gain, that is, a multiplicative constant, and perfect restoration is again achieved.

Random additive noise was handled adequately. This is gratifying since for two of the four Gaussian noise types, the noise values were not all nonnegative. It could be anticipated that the nonnegative nature of the slack variables in linear programming would require a priori knowledge of the most negative noise element to allow a bias to be added. Fortunately, this does not appear to

Edge contrast	$\bar{X} = 0.2$ Case 1	$\sigma = 0.05$ Case 2	Average	$\bar{X} = 1.0$ Case 1	$\sigma = 0.5$ Case 2	Average
10:1	0.79	0.74	0.77	2.36	3.29	2.83
5:1	0.92	0.61	0.77	2.69	2.74	2.72
2:1	1.03	0.90	0.97	2.68	2.52	2.60

Edge contrast	Fourier reconstruction fidelity defect
10:1	5.68
5:1	5.06
2:1	2.48

FIGURE 6.22. Fidelity defect (object [3]).

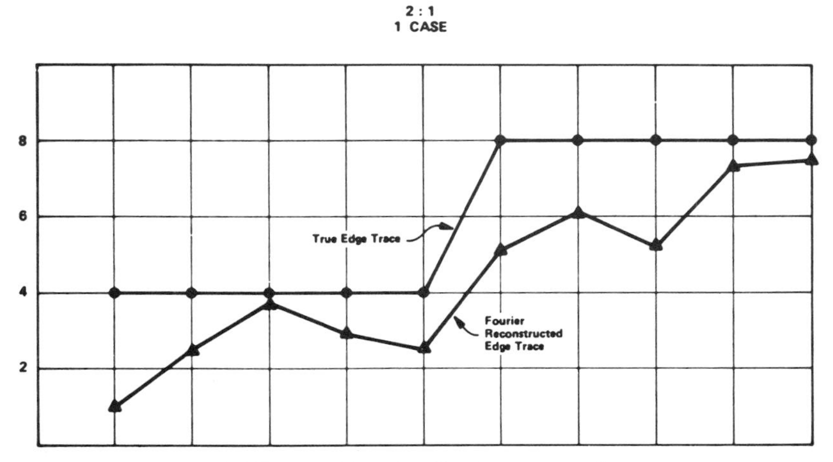

FIGURE 6.23. No noise, Fourier reconstruction.

be critical. If, after further study, negative noise becomes a problem, the optimal bias can be found by directed search.

Several problem areas have shown themselves. The first of these deals with the nature of linear programming. Consider the 2×2 object, 3×3 PSF, and 4×4 image case. This translates into a 16-equation linear programming model with 20 columns (4 object variables, 16 slack or noise variables). Of these 20 variables, only 16 (the number of equations) can be nonzero. It is disconcerting to have to require some number of objects and/or noise variables to be exactly zero. A modification consisting of some additional constraint equations seems indicated.

A second problem is related to the path chosen by the linear programming algorithm from the start point to the final optimal point of the computation. When faced with a choice between two equivalent (as measured by the merit function) paths, the algorithm systematically selects the path involving the lower numbered variable. This leads to a geometric bias, a nonsymmetry in the resulting object and noise arrays. This was overcome by invoking figure of merit (3) in the study, but further consideration of this effect seems warranted.

The increasing complexity of the strategies serves to reduce the calculating time. This is reflected in the number of iterations that are needed to arrive at an optimum solution. In the no-noise case with the first object distribution, 10 iterations are needed, whereas with the fourth strategy, only four are required.

In general, the topic of merit function design, the determination of the optimal PSF, the operational problems associated with larger images, considerations of other types of noise, and application of the technique to real images constitute the logical next areas of study.

In principle, the PSF that produces the best quality measures value can also be computed by parametric variation. It may not be the actual physical PSF, but that is not relevant. Actually, the best estimate of the PSF is used as a starting value, and it can be perturbed within its limit of measurement uncertainly to maximize the figures of merit. This has not been attempted here. Also, the question of computing a different best PSF in different parts of the image is one of several related topics to be addressed in later studies. Finally, the extension to nonlinear figures of merit should be considered.

EDGE DETECTION AND REGIONALIZED TERRAIN CLASSIFICATION FROM SATELLITE PHOTOGRAPHY[5]

Background

The topic of automatic terrain classification from texture as a viable approach to automatic image interpretation has emerged gradually, together

[5]This section was originally published as Hord, R. M., Edge detection and regionalized terrain classification from satellite photography, *Computer Graphics and Image Processing,* 1975, *4,* 184–199.

with classification based on temporal information, after being overshadowed for some time by multispectral classification. Ultimately, a processing system exploiting all three domains, to the extent that these data sources complement one another, may prove to be the most attractive approach to a wide class of problems (Bargel, 1980). The techniques and algorithms reported here ad-

FIGURE 6.24. A Salton Sea–Imperial Valley photograph obtained by Apollo IX (red band).

TABLE 6.15
ITEK Scanner Characteristics

Parameter	Measured limits
Format size, inches	3 × 3
Spot sizes, mils	1,2,4
Points per picture	3,000 ×3,000; 1,490 × 1,490; 740 × 740
Distortion,[a] μm/format	50
Noise, percent full scale	.2
Dynamic range	
Read	100:1
Write	100:1
Quantizing levels (read and write)	1024

[a] Deviation of point from true position.

dress the spatial domain classification problem, that is, automatic interpretation of an image obtained in a single spectral band.

Figure 6.24 is a Salton Sea–Imperial Valley photograph obtained by Apollo IX; the red band is shown. This reversed frame was digitized by the Itek STL Scanner using the 1 mil (25 μm) spot size option and coded with 6 bits (64 gray levels). The scanner characteristics are given in Table 6.15. The goal of this development effort was to develop a software system that would assign grid squares of a satellite-acquired digitized image to land use categories on the basis of spatial information alone. The algorithms were developed in FORTRAN for a CDC 3300 computer using the image in Figure 6.24. The

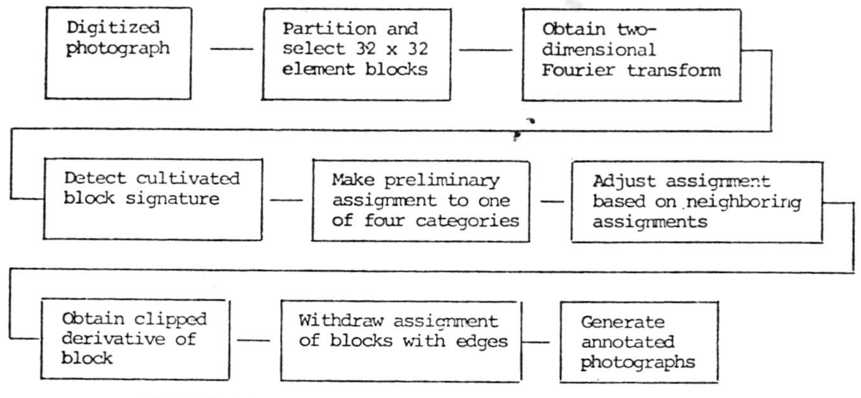

FIGURE 6.25. A block diagram summary chart for the PTCS.

initial categories considered were: (*a*) cultivated land, (*b*) cloud covered terrain, (*c*) water, and (*d*) uncultivated land.

The product of this development effort was the Pattern Terrain Classification System (PTCS). A block diagram summary chart for the PTCS is shown in Figure 6.25. This is a multistage processing sequence that initially assigns each 32 × 32 pixel block of the format provisionally to one of the four categories. This assignment is based on parameters extracted from gray levels exhibited by the blocks in the spatial domain. The 32 × 32 block size permits the efficient use of the FFT algorithm and seems to be about the minimum size for reliable determination of the frequency domain signature. The second processing stage reviews the preliminary assignment matrix from a more global point of view. Assignments are adjusted on the basis of nearby assign-

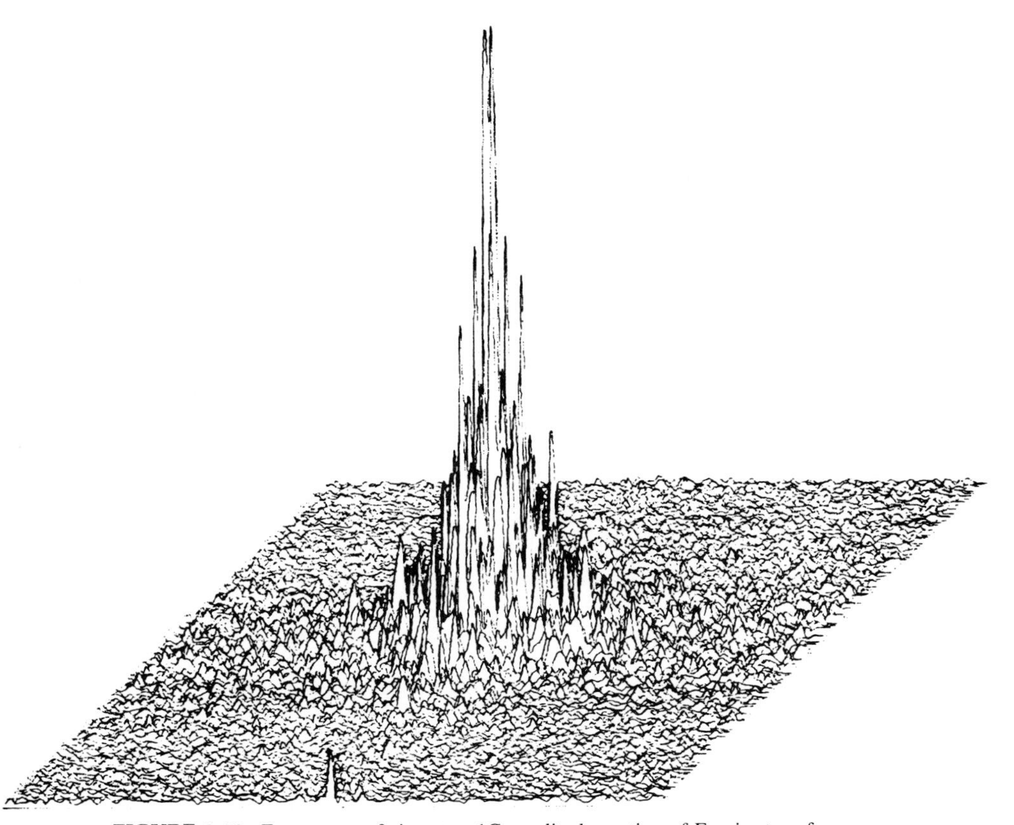

FIGURE 6.26. Emergence of signature 1C amplitude portion of Fourier transform.

ments if the preliminary pattern is improbable. In the third and final stage of processing, the derivative of the block is assessed to screen out those blocks through which a terrain class boundary passes, since to assign such a block to any category would result in a partial misassignment.

The enhancement of the frequency domain signature can be seen in the Figures 6.26–6.29. Figure 6.26 shows a perspective plot of the Fourier transform of a block containing cultivated terrain. The perspective plot of that transform after the pertinent features (the four peaks) have been emphasized, as shown in Figure 6.27. The effect of the same processing on the transform

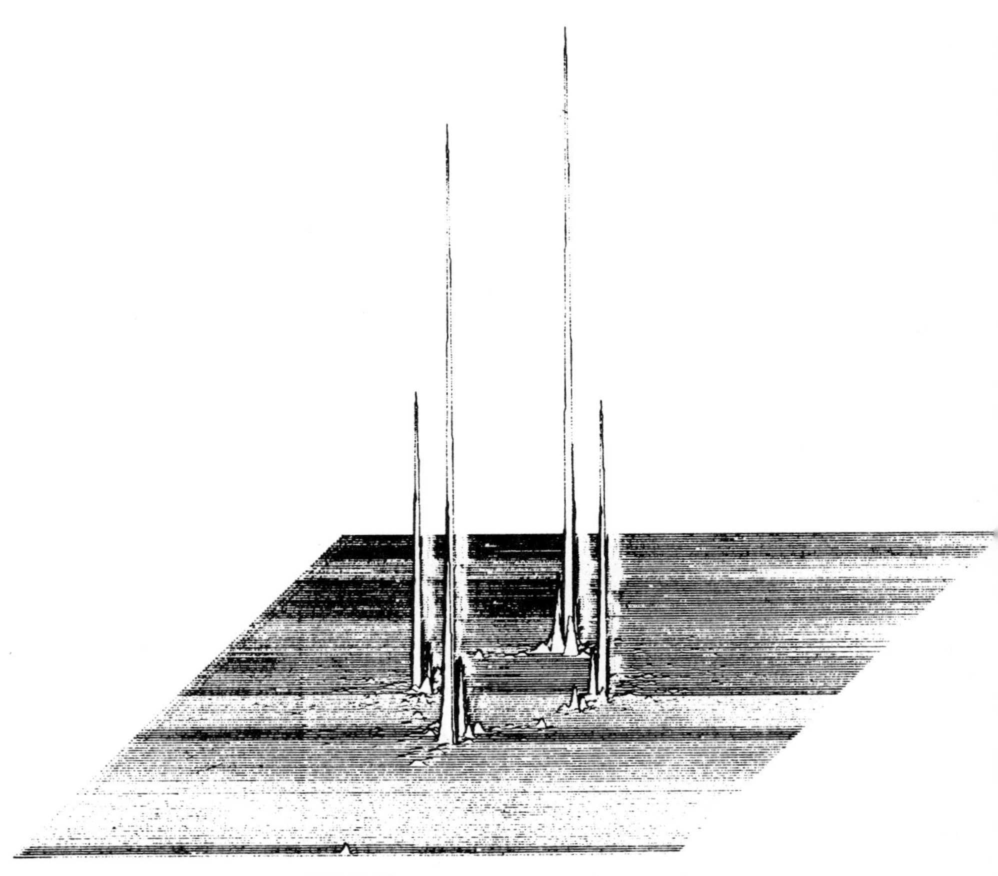

FIGURE 6.27. Signature plot for area 1C.

of a block of uncultivated land is shown in Figure 6.28, illustrating the absence of the characteristic peaks. One may gray-level code the blocks according to the resulting PTCS assignment and record a gray-square display, as given in Figure 6.29, to assess the resolution of the categorization and qualitative success. Alternatively, one may alphabetically code the PTCS assignments and overwrite the results on the original image, as in Figure 6.30.

A comparison matrix relating the PTCS results from Figure 6.30 to the assignments generated by a human photointerpreter is given in Table 6.16; the goodness of fit indicated in this matrix is quantified by $\chi^2 = 2829$. The algorithm that was developed using Figure 6.24 was tested on another data set; the results are shown in Figure 6.31, superimposed on the photograph.

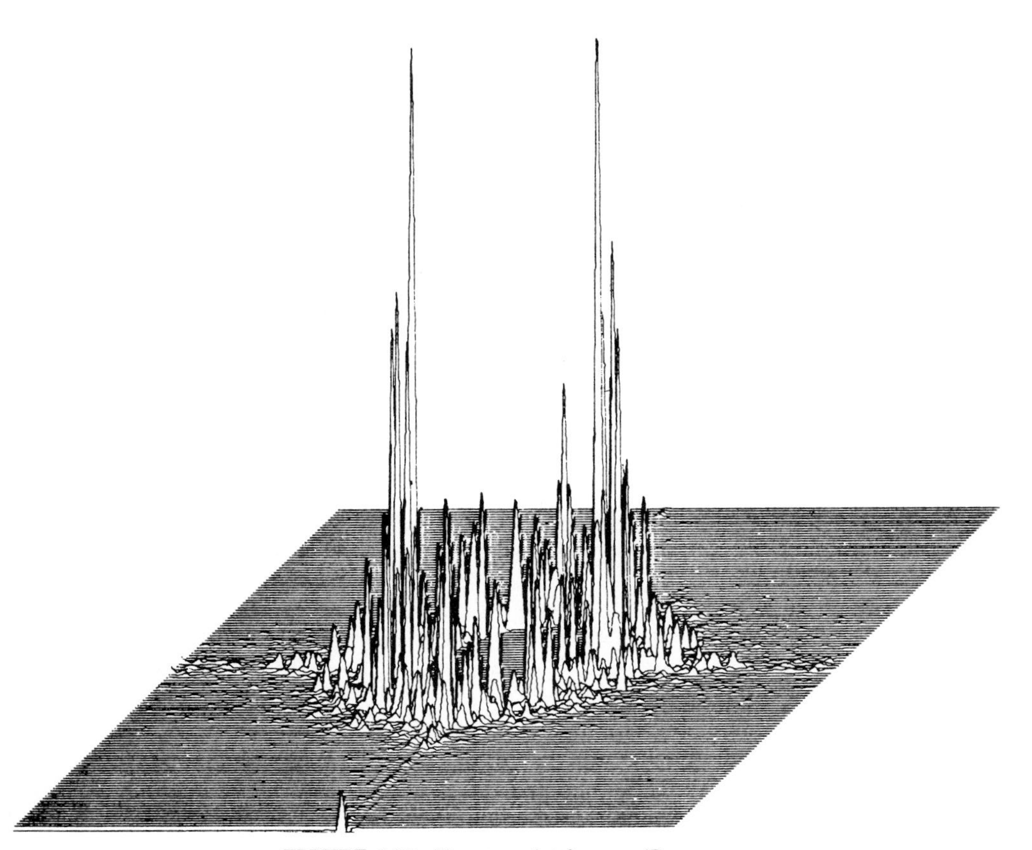

FIGURE 6.28. Signature plot for area 4D.

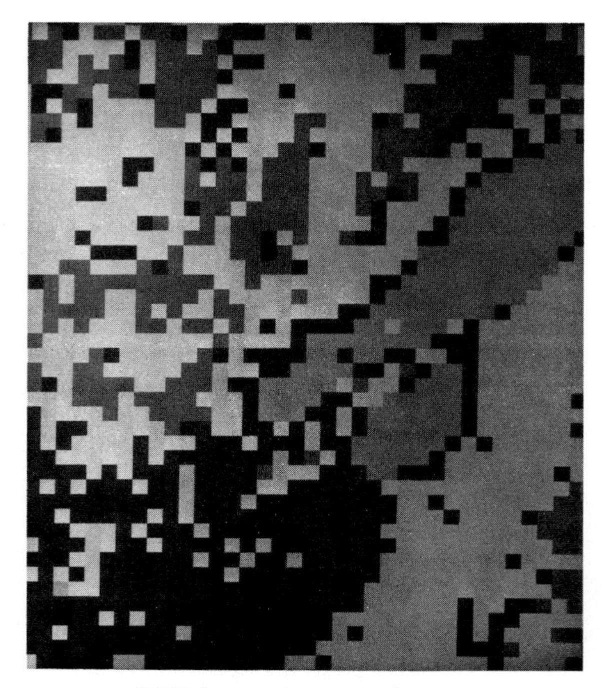

FIGURE 6.29. Gray square display.

The comparison matrix associated with Figure 6.31 is displayed in Table 6.17, characterized by $\chi^2 = 1176$.

Regional Terrain Classification System

In a number of ways these results indicated some promise, but among the deficiencies of the PTCS processing was the heavy dependence on the Fourier transform and the consequent large consumption of computer time. This consideration, together with the desire for an automatic, high-resolution, terrain class boundary map (Hord & Gramenopoulos, 1971), gave rise to the development of the Regional Terrain Classification System (RTCS); the block diagram summarizing the RTCS is given in Figure 6.32.

The central concept of the RTCS is that, before classification, the boundary detection problem is undertaken to partition the format into regions each

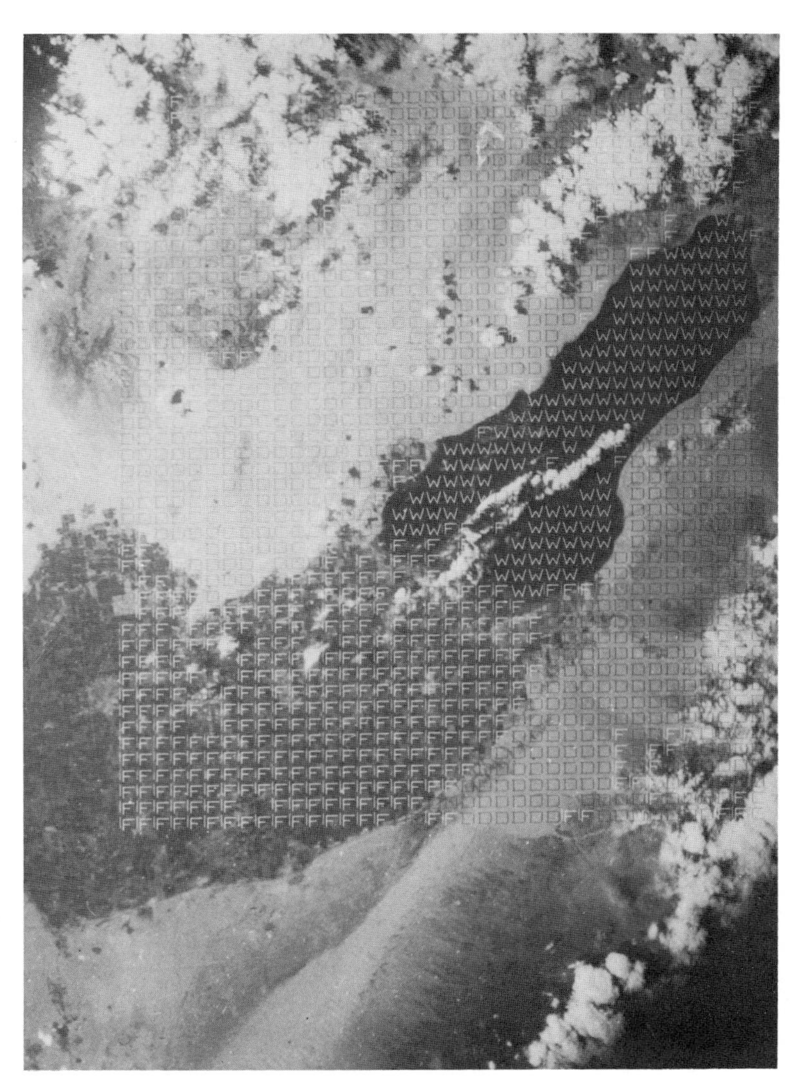

FIGURE 6.30. The Apollo IX photograph, annotated.

TABLE 6.16
Comparison Matrix between Photointerpreter and PTCS

PTCS assignment	Photointerpreter's assignment 1	2	3	4	Total
1	317	19	19	77	432
2	0	145	0	126	271
3	0	0	158	0	158
4	31	61	5	715	812
Total	348	225	182	918	1673

exhibiting a single terrain class (Grinaker, 1980). The regions may then be sampled in the form of several 32 × 32 pixel sets per region, and each of the samples may then be assigned to a terrain class using the techniques employed by the PTCS. If all of the samples of a region agree in assignment, the entire region is classified accordingly; if disagreement is encountered within a region, an omitted boundary is indicated in a specific vicinity. This indication is then employed to improve the boundary map. This boundary map is in low resolution; if a high-resolution map is desired as well as the classification matrix, routines are available for using the low-resolution map as a plan to find boundary detail.

Obtaining Low-Resolution Edge Map

The first operations of the RTCS are a pair of sequential 5× reductions for a net reduction of 25×, and a 1500 × 2250 pixel image then measures 60 × 90. The number of data items is now down by a factor of 625. The reason for the repeated application of a 5× reduction rather than a single 25× reduction is that a 5× reduction is needed later. These reductions are done by averaging and are fairly time consuming. But since, in a production context, this could be performed photographically or by changing the size of the scanning spot, time consumed here is not a source of alarm.

The reduced data set is then subjected to a derivative operation (Rosenfeld, 1980), as described in Figure 6.33. Those pixels for which the threshold criterion is satisfied are termed *E pixels* and constitute the candidate set of edge elements. Edge following is then undertaken. Associated with each

FIGURE 6.31. Another annotated Apollo IX photograph.

member of the set of E pixels is an angle, θ, as defined by the expression in Figure 6.33. At each E pixel a search for the next connecting edge element is performed in the wedge-shaped area defined by $\theta \pm \phi$, where ϕ is a specified wedge half-angle, out to a maximum range. If more than one element is found, the closest is taken. When no next element is found, as when another edge previously followed is encountered, the edge of the format is intercepted or the edge simply trails off; then a new edge following is begun until the E

TABLE 6.17
Picture 2 Comparison Matrix between Photointerpreter and PTCS

PTCS assignment	Photointerpreter's assignment				
	1	2	3	4	Total
1	522	4	0	61	587
2	0	3	0	37	40
3	0	0	0	0	0
4	78	7	0	996	1081
Total	600	14	0	1094	1708

pixel set is exhausted. The results are then cleaned up by removing all edges that terminate other than at the format border or at another edge. The result is termed a *level 3* (low-resolution) *edge map*. As an aside, it can be mentioned that the results can sometimes be improved by preprocessing the E pixel set before initiating any edge following. Such preprocessing might include the logical AND of the set with another set obtained from the same data by an alternative derivative operation, or the systematic elimination of isolated or surrounded members of the set.

It is clear by this point that several parameter values need to be specified by the user of this processing scheme. Choosing the actual values for ϕ, maximum search range, the threshold for the derivative, the binary parameter designating whether or not to eliminate isolated E pixels, etc., would be a difficult task involving multiple trials and the repeated intervention of direct human judgment, were it not for the technique reported in Hord (1971) for quantitatively assessing the degree of match between edge maps. By using a figure of merit, the user or the computer can iteratively adjust the various parameters to achieve the best match to a human-generated map of a portion of the format or to one for the first frame of a series. This human-generated map corresponds to the training set of a supervised classification. The parameters are learned on the training set and then applied to the remainder of the imagery.

Regional Classification

Returning now to the progression shown in the RTCS block diagram of Figure 6.32, the next step is counting the regions that have been delineated. The process is rather simple. Each nonedge pixel's horizontal and vertical

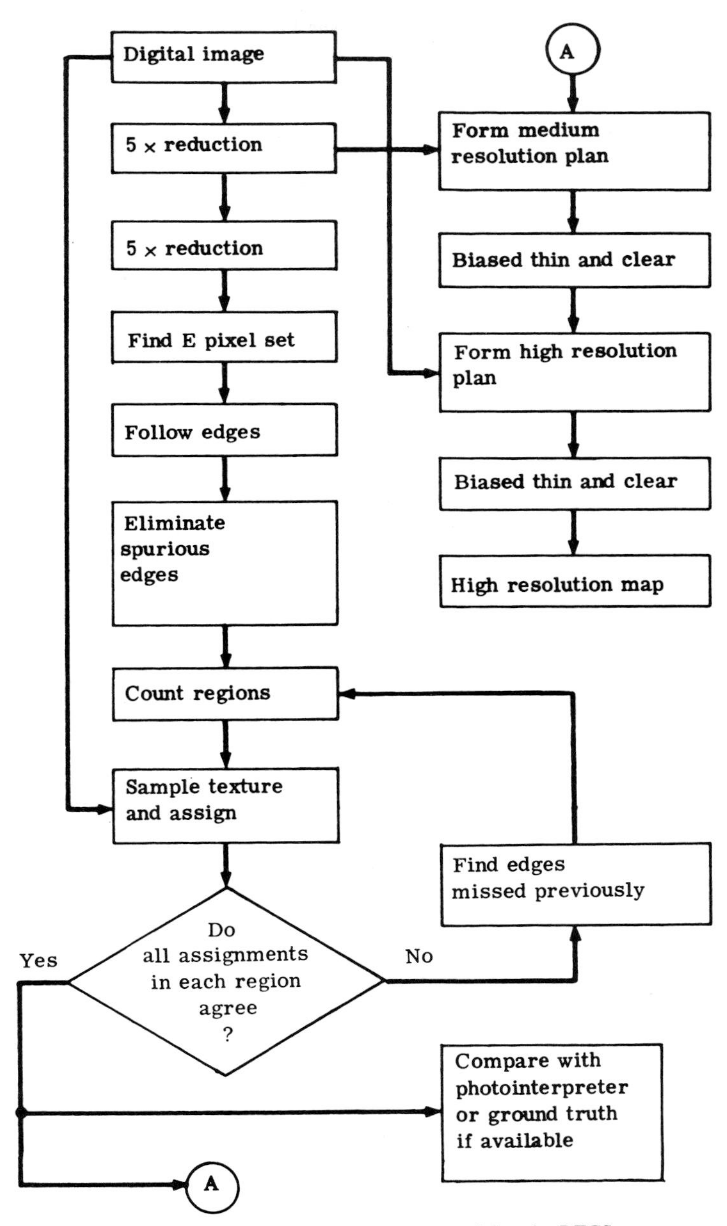

FIGURE 6.32. Block diagram summarizing the RTCS.

$$\frac{(dg/dx)^2 + (dg/dy)^2}{g^2} \geq T$$

where

$$(dg/dx) = g(I + 1, J + 1) + g(I, J + 1) - g(I + 1, J) - g(I, J)$$

$$(dg/dy) = g(I + 1, J + 1) + g(I + 1, J) - g(I, J + 1) - g(I, J)$$

and

$$g(I, J) = G(I, J) + C$$

$$\theta = \tan^{-1}[(dg/dy)/(dg/dx)] + \pi/2$$

FIGURE 6.33. Edge candidate equations.

neighbors are inspected to determine whether any have been assigned to a region number. If none have, the test pixel is assigned the next-larger region number. If some have, and if all neighbor region numbers agree, the test pixel is assigned to that region. If the neighbor region numbers do not agree, the test pixel is given the lowest number, and all members of the disagreement regions are reassigned to that lower region number. After all nonedge pixels have been assigned, the list is collapsed to eliminate region numbers with no pixel assignments.

Sample pixels are then selected and assigned to a terrain class. The conservative sampling scheme employed 25% of all nonedge pixels, uniformly distributed in the format. It is suspected that an 11% sample (one from each 3 × 3 group) would be adequate, but this has not been tested. A pixel can be assigned to a terrain class on the basis of texture since each of these level 3 (low-resolution) pixels corresponds to a 25 × 25 block of level 1 (high-resolution original) pixels. The PTCS classification algorithm described earlier is applied to these square blocks of level 1 pixels (additional level 1 pixels are included in order to provide a properly centered block measuring 32 pixels on a side).

In the pictures processed to date, level 1 formats measuring 1600 × 2250 pixels gave rise to 60 × 90 level 3 formats of which about 400 pixels are edge pixels. Of the 5000 remaining pixels, about 1250 are classified to a terrain category. About 50 regions are found to have an average of 25 classified pixels each.

It is entirely possible that the format partitioning process accepted a spurious edge or omitted a valid edge. The terrain categories of all the samples in a given region are inspected by computer; disagreements flag an omitted edge, and corrective action is taken to include the omitted edge in the level 3 edge map, thereby increasing the region count. On completion of this repair work, each region can be assigned in its entirety to a unique terrain category. In theory, one could at this stage eliminate spurious edges by searching for contingent regions assigned to the same terrain type; In practice, this operation has not been attempted.

Two activities remain to be described: generation of the high-resolution edge map and comparison of the classification results with manual photointerpretation or ground truth. These two operations are independent of one another. To keep the algorithm description unified, the high-resolution edge map generation procedures will be addressed first.

Obtaining High-Resolution Edge Map

A technique reported by Kelly (1970) shows how the concept of planning can be exploited in the context of obtaining a high-resolution edge map. Essentially, the corrected level 3 edge map functions as a plan to direct the search for a medium-resolution (level 2) edge map, which in turn functions as a plan to direct the search for the high-resolution (level 1) edge map. The completed edge map should consist of edges of terrain class regions, and the edges should exhibit the characteristics enumerated in Table 6.18.

In Figure 6.34, a 3×4 pixel portion of a hypothetical level 3 edge map is shown, together with the corresponding level 2 plan. Of the 12 pixels in the level 3 edge map, 5 are edge pixels, designated by *X*s. Each of these corresponds in level 2 to a 5×5 pixel block; the 3×4 pixel level 3 area expands to a 15×20 pixel area in level 2. The plan is then a 15×20 pixel mask with the possible level 2 edge pixels designated there by *X*s. The selected edge pixels will consist of a subset of these possible items. Only pixels in the plan are considered, thereby eliminating a substantial portion of the data set; in practice, about 90% of the level 2 pixels are removed from consideration at this point.

To select the items of the plan to be edge pixels, the level 2 image is used. The mask is encoded with three levels: pixels not in the plan are coded as 0,

TABLE 6.18
Edge Constraints

1. 1 pixel wide
2. Terminate on intersections or format border
3. Continuous
4. Moderate curvature

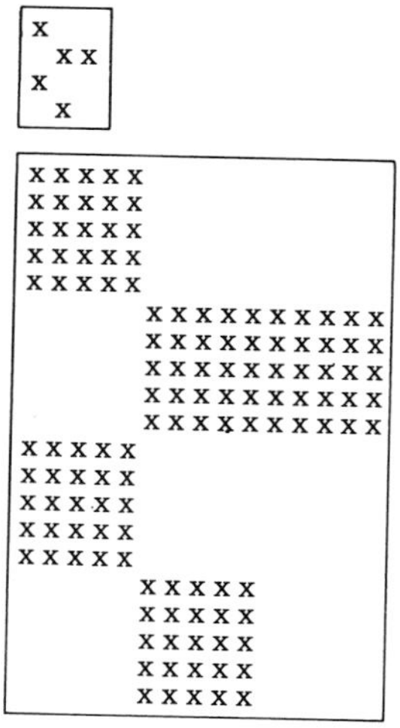

```
x
  x x
x
  x
```

```
x x x x x
x x x x x
x x x x x
x x x x x
x x x x x
          x x x x x x x x x x
          x x x x x x x x x x
          x x x x x x x x x x
          x x x x x x x x x x
          x x x x x x x x x x
x x x x x
x x x x x
x x x x x
x x x x x
x x x x x
          x x x x x
          x x x x x
          x x x x x
          x x x x x
          x x x x x
```

FIGURE 6.34. Edge planning—Part 1: A 3 × 4 pixel portion of a hypothetical level 2 plan.

pixels in the plan are coded with a 1 or a 2. Each 5-element row of a 5 × 5 block in the plan is found in the level 2 image and subjected to a derivative operation, using the same derivative operation as before. The member of the row associated with the largest derivative is assigned a code of 2, and the others are coded with 1s. This gives rise to a ternary coded plan; a hypothetical ternary coded plan is shown in Figure 6.35.

The stage has now been set for generating the level 2 edge maps. The level 2 edges tend to go through the high derivative values, that is, the 2s of the ternary coded plan, but there is no assurance that the 2s, taken as a group, will retain the topological characteristics of the level 3 edge map. To retain the desired continuity, a thinning, or shrinking, algorithm is employed. The algorithm used is based on one described by Stefanelli and Rosenfeld (1971). The major modification of the algorithm is its application to a ternary data set, rather than a binary set. This biases the thinning operation to retain the high derivative valued pixels, and hence the desired edges, without loss of continuity and other topological characteristics. The hypothetical result of applying this biased thin operation to the ternary coded plan of Figure 6.35 is shown as the level 2 edge map in Figure 6.36, together with the level 3 edge map shown earlier. In passing, it should be noted that there is a second modification to the Stefanelli–Rosenfeld algorithm. It consists of a cleanup operation that removes "twigs," that is, short branches that appear along the main edge, terminating at the plan boundary. The procedure, as with level 3,

```
121110000000000
112110000000000
211110000000000
112110000000000
111210000000000
000001211111211
000001121112111
000001211121111
000002111121111
000002111121111
111210000000000
121110000000000
211110000000000
211110000000000
112110000000000
000002111100000
000001121100000
000001111200000
000001112100000
000001112100000
```

FIGURE 6.35. Edge planning—Part 2: A hypothetical ternary coded plan.

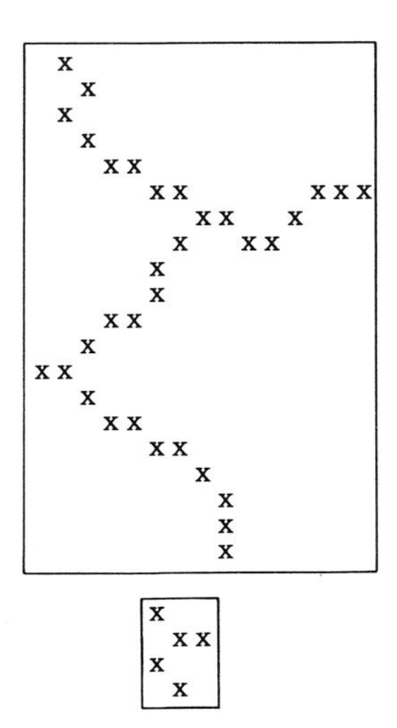

FIGURE 6.36. Edge planning—Part 3: The hypothetical result is this level 2 edge map, with level 3 edge map shown in Figure 6.34.

is to delete any pixels with only one neighbor. This is repeated until no additional deletions occur.

The generation of the level 1 edge map from the level 2 edge map requires an additional application of the same set of procedures. With each application, the total number of pixels in the image increases by a factor of 25, whereas the number of items in the set of pixels increases by a factor of about 5 so that the percentage of the format obscured by edge indicators goes down by a factor of 5 at each stage, for example, from 10% to 2% to .4%.

Results

The assessment of the terrain classification feature of the RTCS has been completed. Table 6.19 displays the comparison of the RTCS results with those of a photointerpreter for the image shown in Figure 6.24. The chi square value associated with these results is 3104 as a measure of the goodness of fit. The RTCS has also been applied to the second image, which was earlier classified by the PTCS. The results of this classification, again compared with that of an interpreter, are shown in Table 6.20. The goodness of fit is here quantified by $\chi^2 = 1719$. In both cases, the chi square value has improved over that obtained by the PTCS software.

If, as is not uncommon, the primary concern is the detection of cultivated land, then the results are perhaps more effectively compared in Table 6.21. There, values of the detection rates and false alarm rates for cultivated land are shown for the two classification systems on the two data sets. Certainly two data sets do not offer conclusive evidence, but the indication is quite promising.

TABLE 6.19
Picture 1 Comparison Matrix between Photointerpreter and RTCS

RTCS assignment	Photointerpreter's assignment				
	1	2	3	4	Total
1	411	3	19	43	476
2	0	92	0	111	203
3	0	4	168	0	172
4	28	56	10	818	912
Total	439	155	197	972	1763

TABLE 6.20
Picture 2 Comparison Matrix between Photointerpreter and RTCS

RTCS assignment	Photointerpreter's assignment 1	2	3	4	Total
1	631	1	0	16	648
2	0	0	0	225	225
3	0	0	0	0	0
4	22	4	0	972	998
Total	653	5	0	1213	1871

TABLE 6.21
Photographs 1 and 2: Cultivated Terrain

	PTCS	RTCS
	Photograph 1	
Detection rate	91.0	94.0
False alarm rate	6.9	3.7
	Photograph 2	
Detection rate	87.0	97.0
False alarm rate	3.8	.9

Conclusion

We began by trying to develop a digital image terrain classification system requiring significantly fewer Fourier transforms than the PTCS scheme to improve the efficiency of the processing. What emerged, or rather, is emerging, is a software system that meets the stated goals and seems to offer some fringe benefits in the form of improved classification results and an automatic delineation of terrain class boundaries. A tabulation of the number and sizes of regions is available, provided the software is improved to delete boundaries separating regions of the same terrain type.

A number of new techniques are embodied in this software, but the key concept is that of planning, which is not new at all. Planning, by attacking a problem in simple stages that systemmatically exclude whole classes of incorrect answers as a group, avoids the inefficiency of brute force approaches. To the digital image processing community, beset with lingering cost-effectiveness problems, planning can be a tool of enormous power if properly exploited.

REFERENCES

Agrawala, A. K., & Kulkarni, A. V. A sequential approach to the extraction of SHAPE features. *Computer Graphics and Image Processing, 1977, 6*(6), 538.

Bargel, B. *Classification of remote sensed data by texture and shape features in different spectral channels.* Paper presented at the IEEE 5th international conference on pattern recognition, Miami Beach, Florida, December 1980.

Barnes, C. W. Object restoration in a diffraction limited imaging system. *Journal of the Optical Society of America,* 1966, 66, 575.

Brousil, J. K., & Smith, D. R. A threshold logic network for SHAPE invariance. *IEEE Transactions on Electronic Computers,* 1967, *EC-16*(6), 818.

Danielsson, P. E. A new shape factor. *Computer Graphics and Image Processing,* 1978, *7*(2), 292.

Dinstein, I., & Silberberg, T. *Shape discrimination with Walsh descriptors.* Paper presented at the IEEE 5th international conference on pattern recognition, Miami Beach, Florida, December 1980.

Frieden, B. R. Restoring with maximum likelihood and maximum entropy. *Journal of the Optical Society of America,* 1972, *62,* 511.

Frieden, B. R. & Burke, J. J. Restoring with maximum entropy II superresolution of photographs of diffraction-blurred impulses. *Journal of the Optical Society of America,* 1972, *62,* 1202.

Gezari, D. Y., Labeyrie, A., & Stachnik, R. V. Speckle interferometry: Diffraction limited measurements of nine stars with the 200-inch telescope. *The Astrophysical Journal,* 1972, *173,* L1–L5.

Grinaker, S. *Edge based segmentation and texture separation.* Paper presented at the IEEE 5th international conference on pattern recognition, Miami Beach, Florida, December 1980.

Harris, J. L. Sr. Resolving power and decision theory. *Journal of the Optical Society of America,* 1964, *54,* 606. (a)

Harris, J. L. Sr. Diffraction and resolving power. *Journal of the Optical Society of America,* 1964, *54,* 931. (b)

Harris, J. L. Sr. Image evaluation and restoration. *Journal of the Optical Society of America,* 1966, *56,* 569.

Hawkins, J. K., *et al.,* Automatic SHAPE detection for programmed terrain classification. *Proceedings of the Society of Photographic Instrumentation Engineers,* 1966, XVI–1.

Helstrom, C. W. Image restoration by the method of least squares. *Journal of the Optical Society of America,* 1967, *57,* 297.

Hord, R. M. *Comparison of edge and boundary detection techniques.* Paper presented at the EIA committee on automatic imagery pattern recognition symposium, Washington, D.C., December, 1971.

Hord, R. M., & Gramenopoulos, N. *Automatic terrain classification from photography.* Paper presented at the SPIE symposium on remote sensing of earth resources and the environment, Palo Alto, California, November 1971.

Hord, R. M., & Sheffield, C. *The POLAR–Cartesian picture transform* (Tech. Rep.) Washington, D.C.: Earth Satellite Co., 1975.

Huang, G. C., Russell, F. D., & Chen, W. H. *Pattern recognition by Mellin transform*. Paper presented at the EIA/AIPR symposium, University of Maryland, April 1975.

Kelly, M. D. Edge detection in pictures by computer using planning. (Stanford, California: Stanford University, 1970.

Knight, C. E. K. Autocorrelation methods to obtain diffraction limited resolution with large telescopes. *The Astrophysical Journal,* 1973, *176,* LA3–LA5.

McGlamery, B. L. Restoration of turbulence degraded images. *Journal of the Optical Society of America,* 1967, *57,* 293.

Mascarenhas, N. Digital image restoration under a regression model—The unconstrained linear equality and inequality constrained approach. *USCIPI Report* 1974, *520.*

Montanari, U. On the optimal detection of curves in noisy pictures. *Communications of the Association for Computing Machinery,* 1971, *14,* 335.

Papoulis, A. Approximations of point spreads of deconvolution. *Journal of the Optical Society of America,* 1972, *62,* 77.

Pavlidis, T. A review of algorithms for SHAPE analysis. *Computer Graphics and Image Processing,* 1978, *7*(2), 243.

Richardson, W. H. Bayesian-based iterative method of image restoration. *Journal of the Optical Society of America,* 1972, *62, 55.*

Rosenfeld, A. Picture processing:1977. Computer Graphics and Image Processing, 1978, 7(2), 211.

Rosenfeld, A. *Quadtrees and pyramids for pattern recognition and image processing*. Paper presented at the IEEE 5th international conference on pattern recognition, Miami Beach, Florida, December 1980.

Rosenfeld, A., & Kak, A. C. *Digital picture processing*. Academic Press, 1976.

Stefanelli, R. S. & Rosenfeld, A. Some parallel thinning algorithms for digital pictures. *Journal of the Association for Computing Machinery,* 1971, *18,* 255–264.

Wong, R. Y., & Hall, E. L. Scene matching with invariant moments. *Computer Graphics and Image Processing,* 1978, *8*(1), 16.

7 Practical Issues

ORGANIZATIONS

One of the practical issues involved with ditital image processing is knowing what organizations are engaged in similar efforts. Lists of some of the governmental, industrial, and academic organizations participating in the digital image processing community follow. These lists were prepared by the Automatic Imagery Pattern Recognition Committee of the Institute of Electrical and Electronics Engineers (IEEE).

Government Listing

Aberdeen Proving Ground
AFCRI
AFIT
Army Behavior and Systems Research Laboratory
Army Corps of Engineers
Army Material Command (MMRC)
Army Surgeon
Army Tank Automotive Command
ARPA
CIA
Department of Defense
Diamond Labs
Edgewood Arsenal
EPA
EROS Data Center
ESSA
ETL
Federal Highway Administration
Florida, State of
Fort Deitrick
Fort Monmouth

Frankford Arsenal
Goddard Institute
Indiana, State of
Institute for Advanced Computation
Intermntn Forest/Range Experimental Station
Kentucky, Commonwealth of
NADC
NASA/AMES
NASA/ERL
NASA/GSFC
NASA/HQ
NASA/JSC
NASA/KSC
NASA/LEWIS
NASA/MSC
NASA/MTF
NASA/WALLOPS
Naval Aerospace Medical Research Laboratory
Naval Air Systems Command
Naval Dental Research Institute
Naval Ordinance Laboratory
Naval Supply System Command
Naval Undersea Center
Naval Underwater System Center
Naval Weapons Laboratory
NBS
NELC
Night Vision Laboratory
NIH
NISC
NOAA/NESS
NRL
NRTSC
NSA
Oak Ridge National Laboratory
Office of Naval Research
Patrick AFB
Postal Service Laboratory
RADC
Redstone Arsenal
South Carolina, State of
State Department
Texas, State of
Transportation, Dept. of
USAR
USDA
USGS
White Sands Missile Range
WPAFB

Industrial Listing

Actron Industries
Adaptronics
Addressograph–Multigraph
Aerospace Corp.
AIL Information Systems
American Can
AMF Electronics Products Co.
Arthur Holt, Inc.
Aydin Controls
Bai Corp.
Ball Brothers Research Corp.
Battells
Bell Labs
Bendix
Boeing
Boeing Computer Services
C–3
Cardion Electronics
Cargill, Inc.
CBS Laboratories
CDC
CSC
Cognos Corp.
Collsman Instrument Co.
Colorado Video, Inc.
Cornell Aeronautical Laboratories (CALSPAN)
Cybernetic R&D Corp.
Dames & Moore, Inc.
Daniel H. Wagner Assoc.
Deere & Col
Development & Resource Transportation Co.
Dicomed
Draper Labs, Inc.
Earth Satellite Corp.
EMR
ESL

E-Systems, Inc.
Exxon Production Research Co.
Fairchild Industries
General Electric
General Motors
General Research Corp.
Goodyear Aerospace Laboratories
Gould, Inc.
Grumman Aerospace
Haskins Laboratories
Honeywell
HRB–Singer
Hughes Research Laboratories
Information & Education Systems, Inc.
Information International, Inc.
INFOTON
ITEK
IBM
Keuffel & Esser
Litton Systems
LNR Communications, Inc.
Lockheed
Marco Scientific
Martin Marietta
MB Associates
McDonnell Aircraft
Mead Technology Laboratories
Mitre
Motorola
MRI
NSC Computing Corp.
North American Science Center
Northrop
Nuclear Defense Research Corp.
Optronics
Pattern Analysis & Recognition Corp.
Perkin Elmer
Photometrics, Inc.
Philco Ford
Planning Research Corp.
Pratt & Whitney
RCA
Radiation, Inc.
Raytheon
Recognition Systems
Robot Research, Inc.
Rockwell International
Sandia Labs
Scope Electronics
Spatial Data Systems
Speech Communication Research Laboratory, Inc.
Sperry Rand
Synectics Corp.
Systems Control, Inc.
SDC
Systems Research Laboratories
Tech Operations
Technology Service Corp.
Texas Instruments
The Analytic Sciences Corp.
3M
Trace, Inc
TRW
Western Electric
Western Union
Westinghouse
Xerox

Academic Listing

Alaska, University of
Arizona, University of
Auburn University (Alabama)
California, University of, at Berkeley
California, University of, at Davis
California, University of, at Irvine
California, University of, at Los Angeles
California, University of, at San Diego
California, University of, at Santa Barbara
California Institute of Technology
California State University
Carnegie–Mellon University (Pennsylvania)
Case Western Reserve University (Ohio)
Columbia University (New York)
Cornell University (New York)
Dartmouth College (New Hampshire)
Drexel University (Pennsylvania)
Duke University (North Carolina)

East Tennesse State University
Florida, University of
Florida Atlantic University
Georgetown University Medical Center (Washington, D.C.)
Georgia Institute of Technology
Harvard School of Public Health (Massachusetts)
Hawaii, University of
Houston, University of (Texas)
Illinois, University of
Iowa, University of
Jet Propulsion Laboratories (California)
Johns Hopkins Applied Physics Lab (Maryland)
Kansas, University of
Kentucky, University of
LARS
Lawrence Livermore Laboratory
Los Alamos Scientific Laboratories (New Mexico)
Louisville, University of (Kentucky)
Loyola University (Illinois)
Maryland, University of
Massachusetts, University of
Massachusetts Institute of Technology
Michigan, University of
Michigan State University
Minnesota, University of
Missouri, University of
Nebraska, University of
New Mexico, University of
New York University
New York, City University of
New York, State University of
North Carolina State University
Ohio State University
Oklahoma, University of
Oregon State University
Pennsylvania, University of
Polytechnic Institute of Brooklyn (New York)
Princeton University (New Jersey)
Purdue University (Indiana)
Rice University (Texas)
Scripps Institute (California)
South Dakota State University
Southeastern Massachusetts University
Southern Calif, University of
Stanford Research Institute (California)
Temple University Medical School (Pennsylvania)
Tennessee, University of
Texas University
Texas A&M University System, The
Texas Christian University
Texas Technological College
Utah, University of
Virginia, University of
Washington, University of
West Virginia, University of
Wisconsin, University of
Woods Hole Oceanographic Institute (Massachusetts)
Yale University (Connecticut)
Yeshiva University (New York)

COURSES AND APPLICATIONS ASSISTANCE

The EROS Data Center operates several Applications Assistance Facilities that maintain microfilm copies of data archived at the center and provide computer terminal inquiry and other capabilities to the center's computer complex. Scientific personnel are available for assistance in applying the data to a variety of resource and environmental problems and for assistance in ordering data from the center.

It is recommended that Applications Assistance Facilities be contacted by phone or mail in advance so that suitable arrangements can be made for a visit. The addresses and telephone numbers of the EROS Applications Assistance Facilities can be obtained from:

EROS Data Center
U.S. Geological Survey
Sioux Falls, South Dakota 57198

EROS Program Office
U.S. Geological Survey
1925 Newton Square East
Reston, Virginia 22090

At the heart of the data center is a central computer complex that controls a database of many millions of images and photographs of the Earth's surface features, performs searches of data on geographic areas of interest, and serves as a management tool for the entire data reproduction process. The computerized data storage and retrieval system is based on a geographic system of latitude and longitude, supplemented by information about image quality, cloud cover, and type of data.

The National Cartographic Information Center (NCIC) is headquartered in the Geological Survey's National Center in Reston, Virginia. It provides a unique service to those customers requiring information on the availability of cartographic data, including multiuse maps, geodetic control, aerial photography, and space imagery. Qualified personnel in the fields of geodesy, photogrammetry, photography, and cartography are ready to help those with specialized needs. NCIC offices with computer links are located at:

National Cartographic Information Center
U.S. Geological Survey
507 National Center, Room 1C107
Reston, Virginia 22092

Mid-Continent Mapping Center
U.S. Geological Survey
1400 Independence Road
Rolla, Missouri 65401

Rocky Mountain Mapping Center
U.S. Geological Survey
Box 25046
Federal Center, Building 25
Denver, Colorado 80225

Western Mapping Center
U.S. Geological Survey
345 Middlefield Road
Menlo Park, California 94025

Guidance in the use of remotely sensed data is available at the EROS Data Center in the form of scheduled training courses and workshops. The scientific teaching staff at the center periodically offers discipline-oriented courses in subjects such as agriculture, forestry, geography, geology, and hydrology. Visitors to the data center may also obtain assistance in the operation of specialized equipment such as densitometers and additive color viewers and in the use of computerized multispectral systems to classify specific phenomena.

The bulk of the training conducted at EDC consists of courses designed to meet the requirements of agencies of the U.S. Department of the Interior and other branches of the federal government. Each course is planned individually, based on the requirements and schedule of the agency involved and the availability of instructors at EDC. The cost of instructional time and materials is borne by the sponsoring agency. Information on the planning of such a course can be obtained through the Chief of the Training and Assistance Section, Applications Branch, U.S. Geological Survey, EROS Data Center.

Two remote sensing workshops are presented at the data center annually, in May and September, for scientists of other nations. Application materials for the International Workshops can be obtained from:

Chief, Office of International Geology
U.S. Geological Survey
National Center (917)
Reston, Virginia 22092

These workshops are introductory in nature and emphasize manual interpretation of Landsat imagery.

Advanced remote sensing courses for foreign scientists are offered by the U.S. Geological Survey's field center in Flagstaff, Arizona. Information on the advanced series can also be obtained from the Office of International Geology (address given previously).

Responsibility for the training of state and local government personnel is assumed by NASA through its Regional Remote Sensing Applications Program. Those interested in the NASA training program should write to one of the following:

Code 902.1
NASA/Goddard Space Flight Center
Greenbelt, Maryland 20771

Code GA Earth Resources Laboratory
Bay St. Louis, Mississippi 39520

Chief, Technology Applications Branch
MS 242-4
NASA/Ames Research Center
Moffett Field, California 94035

Also, several universities in the United States periodically offer short courses or workshops in various aspects of remote sensing, some of which are presented cooperatively with EDC. Information on these can be obtained at:

Remote Sensing
Research Program
Space Sciences Laboratory
University of California
Berkeley, California 94720
(Resources inventory, vegetation analysis, water resources)

Office of Continuing Engineering Education
George Washington University
Washington, D.C. 20052
(Miscellaneous short courses)

Director of Special Programs
Graduate School of Design,
Gund Hall, Harvard University
Cambridge, Massachusetts 02138
(Terrain analysis workshops)

College of Forest, Wildlife and Range Science
University of Idaho
Moscow, Idaho 83843
(Aerial photointerpretation)

Indiana State University Remote Sensing Laboratory
Department of Geography and Geology
Indiana State University
Terre Haute, Indiana 47808
(Introductory courses)

Laboratory for Applications of Remote Sensing
Purdue University
1220 Potter Drive
West Lafayette, Indiana 47906
(Computer processing)

Remote Sensing Program
School of Natural Resources
University of Michigan
Ann Arbor, Michigan 48109
(Vegetation remote sensing workshops)

Director of Continuing Education
South Dakota School of Mines and Technology
Rapid City, South Dakota 57701
(Applications of remote sensing to mineral exploration)

Remote Sensing Institute
South Dakota State University
Brookings, South Dakota 57006
(Visiting Scientist programs)

Manager: Short Course Program
University of Tennessee Space Institute
Tullahoma, Tennessee 37388
(Transportation engineering)

Courses Offered at NASA/Goddard

Both a 2-week course and an additional 1-week course are on remote sensing concepts and applications, with emphasis on computer processing of

Landsat data, and are held at the Eastern Regional Remote Sensing Applications Center (ERRSAC), Goddard Space Flight Center, Greenbelt, Maryland. Additional information on these courses and tentative course outlines may be obtained at:

Code 902.1
Goddard Space Flight Center
Greenbelt, Maryland 20771
Telephone: (301) 982–5084

IMAGE PROCESSING AND INTERPRETATION SERVICES

The following list of organizations performing image processing and interpretation services is distributed by the EROS Data Center, Sioux Falls, South Dakota.

Image Processing and Interpretation Services

Andrew Stancioff, Consultant
P. O. Box 416
Frederick, Maryland 21701

Battelle Corporation
Pacific Northwest Laboratories
Battelle Blvd.
Richland, Washington 99352

Bechtold Satellite Technology Corp.
17137 East Gale Avenue
City of Industry, California 91745

COMARC Design Systems
The Agriculture Bldg.
Embarcadero at Mission
San Francisco, California 94105

Calspan Corporation
P. O. Box 235
Buffalo, New York 14221

Computer Sciences Corporation
8728 Colesville Road
Silver Spring, Maryland 20910

Comtal Corporation
169 North Halstead
Pasadena, California 91107

Control Data Corporation
P. O. Box 1249
Minneapolis, Minnesota 55440

Earth Satellite Corporation
7222 47th Street
Washington, D.C. 20015

Earth Scientists of Americus
1433 Mitchell Trail
Elk Grove Village, Illinois 60007

Ecographics
International Environmental Analysis
P. O. 706
La Jolla, California 92038

Environmental Research Institute
of Michigan
P. O. Box 618
Ann Arbor, Michigan 48107

ESL, Incorporated
495 Java Drive
Sunnyvale, California 94086

Ford Aerospace & Communications Corporation
P. O. Box 58487
Houston, Texas 77058

General Electric Company
5030 Herzel Place
Beltsville, Maryland 20705

Geospectra Corporation
202 East Washington, Suite 504
Ann Arbor, Michigan 48108

Goodyear Aerospace Corporation
Staran Evaluation & Training Facility
Akron, Ohio 44315

HRB—Singer, Inc.
Environmental Analysis Group
Science Park—Box 60
State College, Pennsylvania 16801

International Business Machines
1800 Frederic Pike
Gaithersburg, Maryland 20760

International Imaging Systems
650 North Mary Avenue
Sunnyvale, California 94086

Jet Propulsion Laboratory
4800 Oak Road
Pasadena, California 91103

Lafayette Data Control
P. O. Box 2023
West Lafayette, Indiana 47906

Lockheed Electronics Company
Remote Sensing Applications Lab
1830 Space Park Drive
Houston, Texas 77058

MacDonald—Deitwiler, & Associates Ltd.
10280 Shellbridge Way
Richmond, B.D.
Canada

Mead Technology Laboratory
3481 Dayton–Xenia Road
Dayton, Ohio 45432

Omega Consultants, Inc.
800 East Girard, Suite 516
Denver, Colorado 80231

Oregon State University
Environmental Remote Sensing Aplications Lab
Corvallis, Oregon 97331

Pattern Recognition—Technology and Application
551 Peralta Hills Drive
Anaheim, California 92807

Pennsylvania State University
Office of Remote Sensing of Earth Resources
219 Electrical Eng. West Bldg.
University Park, Pennsylvania 16802

Public Technology, Inc.
1140 Connecticut Avenue, N.W.
Washington, D.C. 20036

Purdue University
Laboratory for the Application of Remote Sensing
West Lafayette, Indiana 47906

Radian Corporation
8500 Shoal Creek Boulevard
Austin, Texas 78758

Remote Sensing Lab
Dept. of Applied Earth Sciences
Stanford University
Stanford, California 94305

Remote Sensing Program
School of Civil and Environmental Engineering
Cornell University
Hollister Hall
Ithaca, New York 14853

Resources Development Associates
1101 N. San Antonio Rd.
Mountain View, California 94043

Seiscom Delta, Inc.
P. O. Box 36798
Houston, Texas 77036

South Dakota State University
Remote Sensing Institute
Brookings, South Dakota 57007

TRW Systems Group
One Space Park
Redondo Beach, California 90278

Texas A & M University
Remote Sensing Center
College Station, Texas 77843

University of California
Space Sciences Laboratory
Berkeley, California 94720

University of Kansas
Space Technology Center
Lawrence, Kansas 66045

University of Maryland
Computer Vision Laboratory
Computer Science Center
College Park, Maryland 20742

University of Southern California
Image Processing Institute
Los Angeles, California 90007

Western Geophysical
Aero Services Division
8100 West Park Drive
Houston, Texas 77036

Appendix: Fourier Transform FORTRAN Subroutine Listing

```
      SUBROUTINE FOURT(DATAR,DATAI,NN,NDIM,IFRWD,ICPLX,WORKR,WORKI)
      DIMENSION DATAR(1),DATAI(1),NN(1),WORKR(1),WORKI(1),IFACT(20)
      IF(NDIM-1)920,1,1
    1 NTOT=1
      DO 2 IDIM=1,NDIM
    2 NTOT=NTOT*NN(IDIM)
      TWOPI=6.283185307
      NP1=1
      DO 910 IDIM=1,NDIM
      N=NN(IDIM)
      NP2=NP1*N
      IF(N-1)920,900,5
    5 M=N
      NTWO=NP1
      IF=1
      IDIV=2
   10 IQUOT=M/IDIV
      IREM=M-IDIV*IQUOT
      IF(IQUOT-IDIV)50,11,11
   11 IF(IREM)20,12,20
   12 NTWO=NTWO+NTWO
      IFACT(IF)=IDIV
      IF=IF+1
      M=IQUOT
      GO TO 10
   20 IDIV=3
      INON2=IF
   30 IQUOT=M/IDIV
      IREM=M-IDIV*IQUOT
      IF(IQUOT-IDIV)60,31,31
   31 IF(IREM) 40,32,40
   32 IFACT(IF)=IDIV
      IF=IF+1
      M=IQUOT
      GO TO 30
   40 IDIV=IDIV+2
      GO TO 30
   50 INON2=IF
      IF(IREM)60,51,60
   51 NTWO=NTWO+NTWO
      GO TO 70
   60 IFACT(IF)=M
   70 NON2P=NP2/NTWO
      ICASE=1
      IFMIN=1
      IF(ICPLX) 100,71,100
   71 ICASE=2
      IF(IDIM-1)72,72,100
   72 ICASE=3
      IF (NTWO-NP1)100,100,73
   73 ICASE=4
      IFMIN=2
      NTWO=NTWO/2
      N=N/2
      NP2=NP2/2
      NTOT=NTOT/2
      I=1
      DO 80 J=1,NTOT
      DATAR(J)=DATAR(I)
      DATAI(J)=DATAR(I+1)
   80 I=I+2
  100 IF( NON2P-1) 101,101,200
  101 NP2HF=NP2/2
```

```
      J=1
      DO 150 I2=1,NP2,NP1
      IF(J-I2)121,130,130
121   I1MAX=I2+NP1-1
      DO 125 I1=I2,I1MAX
      DO 125 I3=I1,NTOT,NP2
      J3=J+I3-I2
      TEMPR=DATAR(I3)
      TEMPI=DATAI(I3)
      DATAR(I3)=DATAR(J3)
      DATAI(I3)=DATAI(J3)
      DATAR(J3)=TEMPR
125   DATAI(J3)=TEMPI
130   M=NP2HF
140   IF(J-M)150,150,141
141   J=J-M
      M=M/2
      IF(M-NP1)150,140,140
150   J=J+M
      GO TO 300
200   DO 270 I1=1,NP1
      DO 270 I3=I1,NTOT,NP2
      J=I3
      DO 260 I=1,N
      IF(ICASE=3)210,220,210
210   WORKR(I)=DATAR(J)
      WORKI(I)=DATAI(J)
      GO TO 240
220   WORKR(I)=DATAR(J)
      WORKI(I)=0.
240   IFP2=NP2
      IF=IFMIN
250   IFP1=IFP2/IFACT(IF)
      J=J+IFP1
      IF(J-I3-IFP2)260,255,255
255   J=J-IFP2
      IFP2=IFP1
      IF=IF+1
      IF(IFP2-NP1)260,260,250
260   CONTINUE
      I2MAX=I3+NP2-NP1
      I=1
      DO 270 I2=I3,I2MAX,NP1
      DATAR(I2)=WORKR(I)
      DATAI(I2)=WORKI(I)
270   I=I+1
300   I1RNG=NP1
      GO TO (302,301,302,302),ICASE
301   I1RNG=NP0*(1+NPREV/2)
302   IF(NTWO-NP1)600,600,303
303   DO 430 I1=1,I1RNG
      IMIN=NP1+I1
      ISTEP=2*NP1
      GO TO 330
310   J=I1
      DO 320 I=IMIN,NTOT,ISTEP
      TEMPR=DATAR(I)
      TEMPI=DATAI(I)
      DATAR(I)=DATAR(J)-TEMPR
      DATAI(I)=DATAI(J)-TEMPI
      DATAR(J)=DATAR(J)+TEMPR
      DATAI(J)=DATAI(J)+TEMPI
320   J=J+ISTEP
```

```
      IMIN=IMIN+IMIN-I1
      ISTEP=ISTEP+ISTEP
330   IF (ISTEP-NTWO)310,310,331
331   IMIN=3*NP1+I1
      ISTEP=4*NP1
      GO TO 420
400   J=IMIN-ISTEP/2
      DO 410 I=IMIN,NTOT,ISTEP
      IF (IFRWD) 401,402,401
401   TEMPR=DATAI(I)
      TEMPI=-DATAR(I)
      GO TO 403
402   TEMPR=-DATAI(I)
      TEMPI=DATAR(I)
403   DATAR(I)=DATAR(J)-TEMPR
      DATAI(I)=DATAI(J)-TEMPI
      DATAR(J)=DATAR(J)+TEMPR
      DATAI(J)=DATAI(J)+TEMPI
410   J=J+ISTEP
      IMIN=IMIN+IMIN-I1
      ISTEP=ISTEP+ISTEP
420   IF(ISTEP-NTWO) 400,400,430
430   CONTINUE
      THETA=-TWOPI/8.
      WSTPR=0.
      WSTPI=-1.
      IF(IFRWD)502,501,502
501   THETA=-THETA
      WSTPI=1.
502   MMAX=8*NP1
      GO TO 540
500   WMINR=COS(THETA)
      WMINI=SIN(THETA)
      WR=WMINR
      WI=WMINI
      MMIN=MMAX/2+NP1
      MSTEP=NP1+NP1
      DO 530 M=MMIN,MMAX,MSTEP
      DO 525 I1=1,I1RNG
      ISTEP=MMAX
      IMIN=M+I1
510   J=IMIN-ISTEP/2
      DO 520 I=IMIN,NTOT,ISTEP
      TEMPR=DATAR(I)*WR-DATAI(I)*WI
      TEMPI=DATAR(I)*WI+DATAI(I)*WR
      DATAR(I)=DATAR(J)-TEMPR
      DATAI(I)=DATAI(J)-TEMPI
      DATAR(J)=DATAR(J)+TEMPR
      DATAI(J)=DATAI(J)+TEMPI
520   J=J+ISTEP
      IMIN=IMIN+IMIN-I1
      ISTEP=ISTEP+ISTEP
      IF(ISTEP-NTWO)510,510,525
525   CONTINUE
      WTEMP=WR*WSTPI
      WR=WR*WSTPR-WI*WSTPI
530   WI=WI*WSTPR+WTEMP
      WSTPR=WMINR
      WSTPI=WMINI
      THETA=THETA/2.
      MMAX=MMAX+MMAX
540   IF(MMAX-NTWO)500,500,600
600   IF(NON2P-1)700,700,601
```

```
601  IFP1=NTWO
     IF=INON2
610  IFP2=IFACT(IF)*IFP1
     THETA=-TWOPI/FLOAT(IFACT(IF))
     IF (IFRWD)612,611,612
611  THETA=-THETA
612  THETM=THETA/FLOAT(IFP1/NP1)
     WSTPR=COS(THETA)
     WSTPI=SIN(THETA)
     WMSTR=COS(THETM)
     WMSTI=SIN(THETM)
     WMINR=1.
     WMINI=0.
     DO 660 J1=1,IFP1,NP1
     I1MAX=J1+I1RNG-1
     DO 650 I1=J1,I1MAX
     DO 650 I3=I1,NTOT,NP2
     I=1
     WR=WMINR
     WI=WMINI
     J2MAX=I3+IFP2-IFP1
     DO 640 J2=I3,J2MAX,IFP1
     TWOWR=WR+WR
     JMIN=I3
     J3MAX=J2+NP2-IFP2
     DO 630 J3=J2,J3MAX,IFP2
     J=JMIN+IFP2-IFP1
     SR=DATAR(J)
     SI=DATAI(J)
     OLDSR=0.
     OLDSI=0.
     J=J-IFP1
620  STMPR=SR
     STMPI=SI
     SR=TWOWR*SR-OLDSR+DATAR(J)
     SI=TWOWR*SI-OLDSI+DATAI(J)
     OLDSR=STMPR
     OLDSI=STMPI
     J=J-IFP1
     IF(J-JMIN)621,621,620
621  WORKR(I)=WR*SR-WI*SI-OLDSR+DATAR(J)
     WORKI(I)=WI*SR+WR*SI-OLDSI+DATAI(J)
630  I=I+1
     WTEMP=WR*WSTPI
     WR=WR*WSTPR-WI*WSTPI
640  WI=WI*WSTPR+WTEMP
     I=1
     DO 650 J2=I3,J2MAX,IFP1
     J3MAX=J2+NP2-IFP2
     DO 650 J3=J2,J3MAX,IFP2
     DATAR(J3)=WORKR(I)
     DATAI(J3)=WORKI(I)
650  I=I+1
     WTEMP=WMINR*WMSTI
     WMINR=WMINR*WMSTR-WMINI*WMSTI
660  WMINI=WMINI*WMSTR+WTEMP
     IF=IF+1
     IFP1=IFP2
     IF(IFP1-NP2)610,700,700
700  GO TO (900,800,900,701),ICASE
701  NHALF=N
     N=N+N
     THETA=-TWOPI/FLOAT(N)
```

```
      IF(IFRWD)703,702,703
  702 THETA=-THETA
  703 WSTPR=COS(THETA)
      WSTPI=SIN(THETA)
      WR=WSTPR
      WI=WSTPI
      IMIN=2
      JMIN=NHALF
      GO TO 725
  710 J=JMIN
      DO 720 I=IMIN,NTOT,NP2
      SUMR=(DATAR(I)+DATAR(J))/2.
      SUMI=(DATAI(I)+DATAI(J))/2.
      DIFR=(DATAR(I)-DATAR(J))/2.
      DIFI=(DATAI(I)-DATAI(J))/2.
      TEMPR=WR*SUMI+WI*DIFR
      TEMPI=WI*SUMI-WR*DIFR
      DATAR(I)=SUMR+TEMPR
      DATAI(I)=DIFI+TEMPI
      DATAR(J)=SUMR-TEMPR
      DATAI(J)=-DIFI+TEMPI
  720 J=J+NP2
      IMIN=IMIN+1
      JMIN=JMIN-1
      WTEMP=WR*WSTPI
      WR=WR*WSTPR-WI*WSTPI
      WI=WI*WSTPR+WTEMP
  725 IF(IMIN-JMIN)710,730,740
  730 IF(IFRWD)731,740,731
  731 DO 735 I=IMIN,NTOT,NP2
  735 DATAI(I)=-DATAI(I)
  740 NP2=NP2+NP2
      NTOT=NTOT+NTOT
      J=NTOT+1
      IMAX=NTOT/2+1
  745 IMIN=IMAX-NHALF
      I=IMIN
      GO TO 755
  750 DATAR(J)=DATAR(I)
      DATAI(J)=-DATAI(I)
  755 I=I+1
      J=J-1
      IF(I-IMAX)750,760,760
  760 DATAR(J)=DATAR(IMIN)-DATAI(IMIN)
      DATAI(J)=0.
      IF(I-J)770,780,780
  765 DATAR(J)=DATAR(I)
      DATAI(J)=DATAI(I)
  770 I=I-1
      J=J-1
      IF(I-IMIN)775,775,765
  775 DATAR(J)=DATAR(IMIN)+DATAI(IMIN)
      DATAI(J)=0.
      IMAX=IMIN
      GO TO 745
  780 DATAR(1)=DATAR(1)+DATAI(1)
      DATAI(1)=0.
      GO TO 900
  800 IF(NPREV-2)900,900,805
  805 DO 860 I3=1,NTOT,NP2
      I2MAX=I3+NP2-NP1
      DO 860 I2=I3,I2MAX,NP1
      IMAX=I2+NP1-1
```

```
      IMIN=I2+I1RNG
      JMAX=I3+I3+NP1-IMIN
      IF(I2-I3)820,820,810
  810 JMAX=JMAX+NP2
  820 IF(IDIM-2)850,850,830
  830 J=JMAX+NP0
      DO 840 I=IMIN,IMAX
      DATAR(I)=DATAR(J)
      DATAI(I)=-DATAI(J)
  840 J=J-1
  850 J=JMAX
      DO 860 I=IMIN,IMAX,NP0
      DATAR(I)=DATAR(J)
      DATAI(I)=-DATAI(J)
  860 J=J-NP0
  900 NP0=NP1
      NP1=NP2
  910 NPREV=N
  920 RETURN
      END
```

Glossary[1]

A

Acutance is a measure of the sharpness of edges in a photograph or image. It is defined for any edge by the average squared rate of change of the density across the edge divided by the total density difference from one side of the edge to the other side of the edge.

An optical color combiner is an instrument that produces false or true color images by linearly combining a few black and white transparencies of the same scene. The transparencies are usually obtained from multispectral, multiband, or time-sequential photography. The transparencies are placed in projectors that are all focused and registered on the same screen and have various color filters placed in front of their lenses. The viewing brightness of the projector's lamp in each projector can be changed independently, thereby changing chromaticity balance. An optical color combiner is sometimes called an **additive color display.**

B

A simple **Bayes decision rule** is one that treats the pixels independently and assigns a pixel having pattern measurements or features d to the category c whose conditional probability, $P_d(c)$, given measurement d, is highest.

C

Each pixel is assumed to be of one and only one given type. The set of types is called the set of pattern classes or **categories,** each type being a particular category. The categories are chosen specifically by the investigator as being the ones of interest to him.

[1]Adapted from Haralick, R. M. Glossary of remotely sensed image pattern recognition concepts. Washington, D.C.: AIPR Committee of Electronic Industries Association, 1973.

A pixel or other unit is said to be recognized, identified, classified, **categorized,** or sorted if the decision rule is able to assign it to some category from the set of given categories. In military applications, there is a definite distinction between *recognize* and *identify*. Here, for a unit to be recognized, the decision rule must be able to assign it to a type of category, the type having included within it many subcategories. For a unit to be identified, the decision rule must be able to assign it not only to a type of category but also to the subcategory of the category type. For example, a small area ground patch that may be recognized as containing trees may be specifically identified as apple trees.

The **Cartesian product** of two sets A and B, denoted by A $\times$ B, is the set of all ordered pairs in which the first component of the pair is some element from the first set and the second component of the pair is some element from the second set. The Cartesian product of N sets can be inductively defined in the usual fashion.

Change detection is the process by which two images may be compared, resolution cell by resolution cell, and an output generated, whenever corresponding resolution cells have different enough grey shades or other characteristic.

Classification is the process in which a decision rule is applied to assign a pixel or other pattern to one of a set of categories.

A **cluster** is a homogeneous group of pixels that are very like one another. Likeness between pixels is usually determined by the association, similarity, or distance between the measurement patterns associated with the pixels.

The pattern classification problem is concerned with constructing the cluster assignment function that groups similar pixels. Pattern classification is synonymous with numerical taxonomy, or **clustering.**

A **cluster assignment function** is a function that assigns each pixel to a cluster on the basis of the measurement pattern(s) in the data sequence or on the basis of their corresponding features. Sometimes the pixels are treated independently; in this case, the clustering assignment function can be considered as a transformation from measurement space to the set of clusters.

A **compound decision rule** is a decision rule that assigns a pixel to a category on the basis of some nontrivial sequence of measurement patterns in the data or in the corresponding feature patterns.

The **conditional probability** of a measurement or feature vector **d** given category c is denoted by $P_c(\mathbf{d})$, or by $P(\mathbf{d}c)$, and is defined as the relative frequency or proportion of times the vector **d** is derived from a pixel whose true category identification is c.

A **confusion matrix** or contingency table is an array of probabilities whose rows and columns are both similarly designated by category label and that indicates the probability of correct identification for each category as well as the probability of type I and type II errors. The (i,k) element P_{ik} *is the probability that a pixel has true category identification* c_i *and is assigned by the decision rule to category* c_k.

Congruencing is the process by which two images of a multi-image set are transformed so that the size and shape of any object on one image is the same as the size and shape of that object on the other image. In other words, when two images are congruenced their geometries are the same and they coincide exactly.

A figure F is **connected** if there is a path between any two spatial coordinates or resolution cells contained in the domain of F.

The **contrast** for a point object against its background can be measured by: (*a*) its **contrast ratio,** which is the ratio between the higher of object transmittance or background transmittance to the lower of object transmittance or background transmittance; (*b*) its **contrast difference,** which is the difference between the higher density of object or background to the lower density of object or background; and (*c*) its **contrast modulation,** which is the difference between the darker of object or background gray shade and the lighter of object or background gray shade divided by the sum of object grey shade and background gray shade.

A figure F is **convex** if the domain of F contains the line segment that joins any pair of spatial coordinates in the domain of F.

D

A **decision boundary** between ith and kth categories is a subset H of patterns in measurement space M defined by

$$H = \{\mathbf{d} \epsilon M | f_i(\mathbf{d}) = f_k(\mathbf{d})\}$$

where f_i and f_k are the discriminant functions for the ith and kth categories.

The **decision rule** f usually assigns one and only one category to each pixel on the basis of the sequence of measurement patterns in the data or in the corresponding feature patterns.

A **densitometer** is a device used to measure the average image density of a small area of specified size on a photographic transparency or print. The measurement may be a meter reading or an electronic signal. When the small area is smaller than a few hundred microns square, the instrument is called a microdensitometer.

Densitometry is the field devoted to the measurement of optical image densities on film or of print gray shades, usually caused by the absorption or reflection of light by developed photographic emulsion.

The **density** of an (x,y) position on a photograph is a measure of the light absorbing capability of the silver or dye deposited on that position. It is defined by the logarithm of the position's reciprocal transmittance. The density measured should be specified as *specular* or *diffuse*.

Provision can be made for the decision rule to reserve judgment or to **defer assignment** if the pattern is too close to the category boundary in measurement or feature space. With this provision, a deferred assignment is an assignment to the category of reserved judgment.

The level slicer, **density slicer,** or thresholder is an instrument (usually electronic) that takes a single or multiple image as an input and produces a binary image for an output. A binary 1 is produced on the output image whenever the gray shades on each of the input images lie within the independently set minimum and maximum thresholds. A set of N input images would, therefore, require a setting of N minimum and N maximum levels.

A pixel or pattern is said to be **detected** if the decision rule is able to assign it as belonging only to some given subset A of categories from the set c of categories. To *detect* a pattern does not imply that the decision rule is able to identify the pattern as specifically belonging to one particular category.

A **digital image, digitized image,** or **digital picture function** of an image is an image in digital format that is obtained by partitioning the area of the image into a finite two-dimensional array of small, uniformly shaped, mutually exclusive regions, called *resolution cells,* and assigning a "representative" gray shade to each such spatial region. A digital image may be abstractly thought of as a function whose domain is the finite two-dimensional set of resolution cells and whose range is the set of gray shades.

A **discrete tonal feature** on a continuous or digital image is a connected set of spatial coordinates or resolution cells, all of which have the same, or almost the same, gray shade.

A **discriminant function** f_i (**d**) is a scalar function whose domain is usually measurement space and whose range is usually the real numbers. When $f_i(\mathbf{d})f_k(\mathbf{d})$ for $k = 1,2, \ldots ,K$, then the decision rule assigns the ith category to the pattern, giving rise to pattern **d.**

A **distribution-free,** or nonparametric, decision rule is one that makes no assumptions about the functional form of the conditional distribution of the patterns given the categories.

E

An **electronic color combiner** is an instrument that produces a false color image by linearly combining video signals of images of the same scenes. The images are usually obtained from multispectral, multiband, or time-sequential photography. If the original image format is photographic, then the image format is changed from photographic to video signal format by synchronized vidicons or flying spot scanners. The resulting video signals are linearly combined through a matrix multiplier circuit, and the three linearly combined signals then drive the color gun of a color television tube. An electronic color combiner usually has greater versatility for congruencing or registering than does an optical color combiner.

A false identification, or false alarm, or type II error occurs for category c_i if a pattern whose true category identification is c_k, $k \neq i$, is assigned by the decision rule to category c_i. A false identification error is often called an **error of commission.**

A misidentification, or misdetection, or type I error occurs for category c_i if a pattern whose true category identification is c_i is assigned by the decision rule to category c_k, $k \neq i$. A misidentification error is often called an **error of omission.**

F

A **feature, feature pattern, feature vector,** or pattern feature is a vector with a small number of components that are functions of the initial measurement pattern variables or some subsequence of the measurement vectors. Feature vectors are designed to contain a high amount of information relative to the discrimination between patterns of the types of categories in the given category set. Sometimes the features are predetermined and other times they are determined at the time the pattern discrimination problem is being solved. In image pattern recognition, features often contain information relative to gray shade, texture, shape, or context.

Feature extraction is the process in which an initial measurement pattern or some subsequence of measurement patterns is transformed to a new pattern feature. Sometimes feature extraction is called *property extraction.*

Feature selection is the process by which the features to be used in the pattern recognition problem are determined. Sometimes feature selection is called *property selection.*

Feature space is the set of all possible feature vectors.

A **flying spot scanner** is a device used to rapidly convert image data from photographic format to electronic video signal format. Normally, the scanner directs an electron beam across the face of a cathode ray tube (CRT) in a television-like raster. The photographic transparency is placed in front of the CRT (either directly or through some optics) and the light coming from the CRT is passed through it. The modulated light beam is detected by a photomultiplier or other photo detector and amplified to a usable video signal level.

G

The *gray shade* or *gray tone* is a number or value assigned to a position (x, y) on an image. The number is proportional to the integrated output, reflectance, or transmittance of a small area, usually called a *resolution cell* or *pixel,* centered on the position (x,y). The gray shade can be measured as or expressed in any one of the following ways:

1. Transmittance
2. Reflectance
3. A coordinate of the International Color Institute (ICI) color coordinate system
4. A coordinate of the tristimulus value color coordinate system
5. Brightness
6. Radiance
7. Luminence
8. Density
9. Voltage
10. Current

H

A **hyperplane decision boundary** is the special name given to decision boundaries arising from the use of linear discriminant functions.

I

An **image** is a spatial representation of an object, scene, or another image. It can be real or virtual, as in optics. In pattern recognition, *image* usually means a recorded image such as a photograph, map, or picture. It may be abstractly thought of as a continuous function I of two variables defined on some bounded region of a plane. When the image is a photograph, the range of the function I is the set of gray shades usually considered to be normalized to the interval (0,1). The gray shade located at spatial coordinate (x,y) is denoted by $I\ (x,y)$ and is usually proportional to the radiant energy in the electromagnetic band to which the photographic sensor is sensitive. When the image is a map, the range of the function I is a set of symbols or colors, and the symbol or color located at spatial coordinate (x,y) is denoted by $I\ (x,y)$. A recorded image may be in photographic, video signal, or digital format.

Image compression is an operation that preserves all or most of the information in the image and reduces the amount of memory needed to store an image or the time needed to transmit an image.

Image enhancement is any one of a group of operations that improve the detectability of the targets or categories. These operations include, but are not limited to, contrast improvement, edge enhancement, spatial filtering, noise suppression, image smoothing, and image sharpening.

Image processing encompasses all the various operations that can be applied to photographic or image data. These include, but are not limited to, image compression, image restoration, image enhancement, preprocessing, quantization, spatial filtering, and other image pattern recognition techniques.

Image restoration is a process by which a degraded image is restored to its original condition. Image restoration is possible only to the extent that the degradation transform is mathematically invertible.

An **image transformation** is a function or operator that takes an image for its input and produces an image for its output. The domain of the transform operator is often called the *spatial domain*. The range of the transform operator is often called the *transformed domain*. Some transformations have spatial and transform domains of entirely different character. For these transforms, the image in the spatial domain may appear entirely different from and have a different interpretation than the image in the transformed domain. Specific examples of these kinds of transformations are the Fourier and Hadamard transformations. Other transformations have spatial and transform domain of same character. For these transformations, the image in the transformed domain may appear similar to the image in the spatial domain. These types of transformations are often called *spatial filters*.

Interactive image processing refers to the use of an operator or analyst at a console with a means of accessing, preprocessing, feature extracting, classifying, identifying, and displaying the original imagery or the processed imagery for his or her subjective evaluation and further interactions.

L

Level slicing or density slicing or thresholding is an operation performed by an instrument (usually electronic) called a *level slicer* to change one or more gray-scale images to one binary image. A binary image is produced whenever the gray shade is less than (greater than) a specified threshold.

The spread function of an image system, process, component, or material describes the resulting spatial distribution of gray shade when the input to the system is some well-defined object much smaller than the width of the spread function. If the input to the system is a line, the spread function is called the **line-spread function.** If the input to the system is a point, the spread function is called the *point-spread function*.

A **linear decision rule** is a simple decision rule that usually treats the patterns independently and makes the category assignments using linear discriminant functions. The decision boundaries obtained from linear decision rules are hyperplanes.

A **linear spatial filter** is a spatial filter for which the gray shade assignment at coordinates (x,y) in the transformed image is made by some weighted average (linear combination) of gray shades located in a particular spatial pattern around coordinates (x,y) of the domain image. The linear spatial filter is often used to change the spatial frequency characteristics of the image. For example, a linear spatial filter that emphasizes high spatial frequencies will tend to sharpen the edges in an image. A linear spatial filter that emphasizes the low spatial frequencies will tend to blur the image and reduce salt and pepper noise.

A pattern is said to be **located** if specific coordinates can be given for the pattern's physical location.

M

A **map** is a representation of physical and/or cultural features (natural, artificial, or both) of a region (such as the sky) or a surface such as that of the Earth or another planet. It indicates, by a combination of symbols and colors, those regions having designated category identifications. Very often ground truth and/or decision rule category assignments are displayed by maps. A photograph with limited symbology and annotation is often called a *photomap*.

Matched filtering is a template matching operation done by using the magnitude of the cross-correlation function to measure the degree of matching.

A simple **maximum likelihood decision rule** is one that treats the pixels independently and assigns a pixel having pattern measurements or features d to that category c whose units most probably gave rise to pattern or feature vector **d**, that is, such that the conditional probability of **d** given c, $P_c(\mathbf{d})$, is highest.

A **measurement pattern,** or **measurement vector,** is the ordered vector of measurements obtained of a pattern under observation. Each component of the vector is a measurement of a particular quality, property, feature, or characteristic of the pattern. In image pattern recognition, the patterns are usually picture elements or simple formations of picture elements and the measurement vectors are the corresponding gray shades.

Measurement space is a set large enough to include in it the set of all possible measurement vectors that could be obtained by observing physical attributes of some set of units. When the units are single resolution cells or picture elements, measurement space M is the Cartesian product of the range sets of the sensors:

$$M = R_1 \times R_2 \times \cdots \times R_n.$$

The **modulation transfer function** of an imaging system or component measures the spatial frequency modulation response of the system or component. As an imaging system or component processes or records an image, the contrast modulation of the processed or recorded image is different from the input image. In fact, there is always a spatial frequency beyond which the contrast modulation of the processed or recorded (output) image is smaller (worse) than the contrast modulation of the input image. The modulation transfer function can be thought of as a curve indicating, for each spatial frequency, the ratio of the contrast modulation of the output image to the contrast modulation of the input image. It is formally defined as the magnitude of the Fourier transform of the line-spread function of the imaging system or component.

A **multidigital image** is a multiimage in digital form. It can be, for example, a set of digital images obtained from the images in a multiimage. A multidigital image is often called a *multi-image* for short when it is understood from the context that digital images are involved.

A **multiimage** is a set of images, each taken of the same subject at different times, or from different positions, or with different sensors, or at different electromagnetic frequencies, or with different polarizations. Although there is a high degree of information redundancy between images in a multiimage set, each image usually has information not available in any one of or combinations of the other images in the set.

N

A simple **nearest neighbor decision rule** is one that treats the pixels independently and assigns a pixel of unknown identification and with pattern measurements or features d to category c_j where d_j is that pattern closest to d by some given metric or distance function.

P

The word **pattern** can be used in three distinct senses:

1. As measurement pattern
2. As feature pattern
3. As the dependency pattern or patterns of relationships among the components of any measurement vector or feature vector that are derived from units of a particular category and are unique to those vectors

The **pattern discrimination** problem is concerned with how to construct the decision rule that assigns a unit to a particular category on the basis of the measurement pattern(s) in the data sequence or on the basis of the feature pattern(s) in the data sequence.

Pattern recognition is concerned with, but not limited to, problems of: (*a*) pattern discrimination, (*b*) pattern classification, (*c*) feature selection, (*d*) pattern identification, (*e*) cluster identification, (*f*) feature extraction, (*g*) preprocessing, (*h*) filtering, (*i*) enhancement, (*j*) pattern segmentation, and (*k*) screening.

In pattern recognition problems such as target discrimination, for which the category of interest is some specified formation of resolution cells with characteristic shape or tone–texture composition, the problem of pattern segmentation may occur. **Pattern segmentation** is the problem of determining which regions or areas in the image constitute the patterns of interest, that is, which resolution cells should be included and which excluded from the pattern measurements.

A **photograph** is a hard copy pictorial record of an image formed by a sensor. The photograph is usually recorded on some type of photosensitive emulsion. It can be either reflective, as is a paper print, or transmissive, as is a transparency. It is usually two-dimensional, and its reflectance or transmittance (either monochromatic or polychromatic) varies as a function of position. If it is a multicolored image (polychromatic), it can be either natural color where the colors are similar to the original, or false color where the colors of the photograph are radically different from the original. The sensor used to form the image may be any type, such as an optical camera with or without spectral filtration, infrared optical-mechanical scanners, television systems, radars, sonic sensors, etc. The type of sensor recording the image and spectral region to which the sensor is sensitive should always be indicated when referring to a photograph.

A **picture element, pixel,** or **pel** is a pair whose first member is a resolution cell and whose second member is the gray shade assigned by the digital image to that resolution cell. Sometimes *picture element, pixel,* or *pel* refer only to the gray shade or gray-shade vector in a resolution cell.

Preprocessing is an operation applied before pattern identification is performed. Preprocessing produces, for the categories of interest, pattern features that tend to be invariant under changes such as translation, rotation, scale, illumination levels, and noise. In essence, preprocessing converts the measurement patterns to a form that allows a simplification in the decision rule. Preprocessing can bring into registration, bring into congruence, remove noise, enhance images, segment target patterns, detect, center, and normalize targets of interest.

Q

A **quantizer** is an instrument that does quantizing. The quantizer has three functional parts. The first part allows the determining and/or setting of the quantizing intervals; the second part is a level slicer that indicates when a signal is in any quantizing interval; and the third part takes the binary output from the level slicers and either codes it to some binary code or converts it to some analog signal representing quantizing interval centers or means.

Quantizing is the process by which each gray shade in an image of photographic, video, or digital format is assigned a new value from a given finite set of gray-shade values. There are three often-used methods of quantizing:

1. In *equal interval quantizing,* or *linear quantizing,* the range of gray shades from maximum gray shade is divided into contiguous intervals, each of equal length, and each gray shade is assigned to the quantized class that corresponds to the interval within which it lies.

2. In *equal probability quantizing,* the range of gray shades is divided into contiguous intervals such that, after the gray shades are assigned to their quantized class, there is an equal frequency of occurrencefor each quantized gray shade in the quantized digital image or photograph; equal probability quantizing is sometimes called *contrast stretching.*
3. In *minimum variance quantizing,* the range of gray shades is divided into contiguous intervals such that the weighted sum of the variance of the quantized intervals is minimized. The weights are usually chosen to be the gray-shade interval probabilities, which are computed as the proportional area on the photograph or digital image that have gray shades in the given interval.

R

The **radiant intensity** of a point object is a measure of the radiant power per steradian radiated or reflected by an object. In general, radiant intensity is a function of the nature of the object, the viewing angle, spectral wavelength and bandwidth.

Target identification or **recognition** is the process by which targets contained within image data are identified by means of a decision rule.

Rectifying is a process by which the geometry of an image area is made plainmetric. For example, if the image is taken of an equally spaced rectangular grid pattern, then the rectified image will be an image of an equally spaced rectangular grid pattern. Rectification does not remove relief distortion.

The **reflectance,** or **reflection coefficient,** is the ratio of the energy per unit time per unit area (radiant power density) reflected by the object to the energy per unit time per unit area incident on the object. In general, reflectance is a function of the incident angle of the energy, viewing angle of the sensor, spectral wavelength and bandwidth, and the nature of the object.

Registering is the translation–rotation alignment process by which two images of like geometries and of the same set of objects are positioned coincident with respect to one another so that corresponding elements of the same ground area appear in the same place on the registered images. In this manner, the corresponding gray shades of the two images at any (x,y) coordinate or resolution cell will represent the sensor output for the same object over the full image frame being registered.

Resolution is a generic term that describes how well a system, process, component, material, or image can reproduce an isolated object or separate closely spaced object or lines. The *limiting resolution, resolution limit,* or *spatial resolution* is described in terms of the smallest dimension of the target or object that can just be discriminated or observed. Resolution may be a function of object contrast and of spatial position as well as element shape (single point, number of points in a cluster, continuum, or line, etc.)

A **resolution cell** is the smallest, most elementary areal constituent of gray shades considered by an investigator in an image. A resolution cell is referenced by its spatial coordinates. The resolution cell, or formations of resolution cells, can sometimes constitute the basic unit for pattern recognition of image format data.

The **resolving power** of an imaging system, process, component, or material is a measure of its ability to image closely spaced objects. The most common practice in measuring resolving power is to image a resolving power target composed of lines and spaces of equal width. Resolving power is usually measured at the image plane in line pairs per millimeter, that is, the greatest number of lines and spaces per millimeter that can just be recognized. This threshold is usually determined by using a series of targets of decreasing size and basing the measurement on the smallest one in which all lines can be counted. In measuring resolving power, the nature of the target (number of lines and their aspect ratio), its contrast, and the criteria for determining the limiting resolving power, *must be* specified.

S

A **scanning densitometer** is a device used to convert image data from transparency photographic format to electronic video-signal format. Usually, the photographic transparency is placed on a glass cylinder that rotates and slowly translates. A fine beam of light is focused on the transparency, passed through it, and is detected by a photomultiplier, which amplifies it to a usable video signal. The scanning densitometer is a much slower conversion device than the flying spot scanner. However, this disadvantage is compensated by its fine resolution capability (a few microns).

Screening is the operation of separating the uninteresting photographs or images from those photographs containing areas of potential interest.

A **signature** is the observable or characteristic measurement or feature pattern derived from units of a particular category. A category is said to have a signature only if the characteristic pattern is highly representative of the n-tuples obtained from units of that category. Sometimes a signature is called a *prototype pattern.*

A **simple decision rule** is a decision rule that assigns a unit to a category solely on the basis of the measurements or features associated with the unit. Hence, the units are treated independently and the decision rule f may be thought of as a function that assigns one and only one category to each pattern in measurement space or to each feature in feature space.

A **spatial filter** is an image transformation, usually a one–one operator used to lessen noise or to enhance certain characteristics of the image. For any particular (x,y) coordinate on the transformed image, the spatial filter assigns a gray shade on the basis of the gray shades of a particular spatial pattern near the coordinates (x,y).

A **subimage** F in a continuous or digital image I is any function F whose domain is some subset A of the set of spatial coordinates or resolution cells, whose range is the set, G, of grey shades, and which is defined by $F(x,y) = I(x,y)$ for any (x,y) belonging to A. Also called a *window*.

T

A **target** is one type of category used in the pattern recognition of image data. It usually occupies some relatively small area on the image and has a unique or characteristic set of attributes. It has a high priority interest to the investigator.

Target discrimination is the process by which decision rules for targets (small area extensive categories) are constructed.

Template matching is an operation that can be used to find out how well two photographs or images match one another. This degree of matching is often determined by cross-correlating the two images or by evaluating the sum of the squared corresponding gray-shade differences. Template matching can also be used to best match a measurement pattern with a prototype pattern.

Texture is concerned with the spatial distribution of the gray shades and discrete tonal features. When a small area of the image has little variation of discrete tonal features, the dominant property of that area is gray shade. When a small area has wide variation of discrete tonal features, the dominant property of that area is texture. There are three things crucial in this distinction: (*a*) the size of the small areas, (*b*) the relative sizes of the discrete tonal features, and (*c*) the number of distinguishable discrete tonal features.

A **training sequence** is a set of two sequences: (*a*) the data sequence and (*b*) a corresponding category identification sequence (sometimes called *ground truth*). The training sequence is used to estimate the category conditional probability distributions from which the decision rule is constructed.

The **transmittance,** or **transmittance coefficient,** is the ratio of the energy per unit time per unit area (radiant power density) transmitted through the object to the energy per unit time per unit area incident on the object. In general, transmittance is a function of the incident angle of the energy, viewing angle of the sensor, spectral wavelength and bandwidth, and the nature of the object.

U

The simplest and most practical **unit** to observe and measure in the pattern recognition of image data is often the basic picture element (the gray shade or the gray shade vector n-tuple in its particular resolution cell). Pattern recognition is difficult sometimes since the objects requiring analysis or identification can be not just simple picture elements but complex spatial formations of picture elements such as houses, roads, forest, etc.

V

A **video image** is an image in electronic signal format capable of being displayed on the CRT screen. The video signal is generated from devices like a vidicon or a flying spot scanner, which converts an image from photographic form to video signal form by scanning it line by line.

The **vidicon** is an imaging vacuum tube with a photosensitive surface that can be scanned by an electron beam generating a signal whose amplitude corresponds to the radiant intensity focused on the surface at each point. This signal is called a *video signal* and may be amplified to any desired level.

Related Reading

Aggarwal, J. K., Duda, R. O. & Rosenfeld, A. (eds.). Computer methods in image analysis. New York: IEEE Press, 1977.

American Elsevier Publishing Co., *Remote Sensing of Environment.* Quarterly journal. 52 Vanderbilt Avenue, New York, New York 10017.

American Society of Photogrammetry. *Operational Remote Sensing,* 1972. 105 North Virginia Street, Falls Church, Virginia 22046.

—*Proceedings of the Annual Fall Meetings*

—*Proceedings of the Annual March Meetings.*

—1973, Third Biennial Workshop. *Color Photography in the Plant Sciences and Related Fields.*

—*Photogrammetric Engineering and Remote Sensing.* Monthly journal.

Avery, T. E. *Interpretation of aerial photographs. (2nd ed.).* Minneapolis, Minnesota, Burgess, 1968.

Baumann, L. S. (Ed.). Proceedings: Image Understanding Workshop (20 April 1977, Minneapolis Minnesota), Arlington, Virginia: Science Applications, 1977. (a)

Baumann, L. S. (Ed.). *Proceedings: Image Understanding Workshop* 20–21 October 1977, Palo Alto, California. Arlington, Science Applications, 1977. (b)

Colvocoresses, A. P. *Remote sensing platforms.* (U.S. Geol. Circ. 693), 1975.

Colwell, R. N. *Monitoring earth resources from aircraft and Spacecraft.* Washington, D.C.: US Gov. Printing Office, 1971.

Duda, R., & Hart, P. E. *Pattern classification and scene analysis.* New York: Wiley, 1973.

Earth Resources Technology Satellite U.S. Standard Catalog and Earth Resources Technology Satellite Non-U.S. Geological Survey, Sioux Falls, South Dakota 57198. Cumulative catalogs have been published to cover data available from launch of Landsat–1 through 23

July 1974 for the U.S. *Standard Catalog* and from launch of Landsat–1 through 23 July 1973 for the *Non-U.S. Standard Catalog*.

Environmental Research Institute of Michigan. *Proceedings of the Annual International Symposium on Remote Sensing of the Environment*, 1974.P. O. Box 618, Ann Arbor, Michigan 48107.

Estes, J., & Senger, L. (Eds.) *Remote sensing techniques for environmental analysis*. Santa Barbara, California: Hamilton Publishing Co, 1974.

Evans, J. M., Jr., & Kirsch, R., & Nagel, R. N. (Eds.). *Workshop on standards for image pattern recognition* (National Bureau of Standards, Special Publication 500-S). Washington, DC: US Govt. Printing Office, 1977. (Conference held 3–4 June 1976, Gaithersburg, Maryland)

5th International Joint Conference on Articicial Intelligence—1977 (IJCAI-77), *Proceedings of the Conference* (22–25 August 1977, Cambridge, Mass.)

Gonzalez, R. C., & Wintz, P. A. *Digital image processing*. Reading, Massachusetts: Addison Wesley, 1977.

Henderson, R. G., & Lemmon, A. V. *A study of digital image analysis systems* (MTR 7446), McLean, Virginia: MITRE, 1977.

Klinger, A., Fu, K. S., & Kunii, T. L. (Eds.). *Data structures, computer graphics, and pattern recognition*. New York: Academic Press, 1977. (Conference held 14–16, May 1975, Los Angeles, California.)

Meisel, W. S. *Computer oriented approaches to pattern recognition*. New York: Academic Press, 1972.

National Academy of Sciences. *Remote sensing with special reference to agriculture and forestry*. Washington, D.C.: Author, 1970.

National Aeronautics and Space Administration. *Advanced scanners and imaging systems for Earth observations*. Washington, D.C.: NASA–SP 355. (NTIS order no. N74-11287).

—*Remote sensing of Earth resources: A literature survey with indexes*. (NTIS order no. NASA SP–7036 for period 1962–1970; NASA SP–7036(1) for period 1970–1973; NASA SP–7041(1) beginning quarterly in 1974).

—*Symposium of significant results obtained from Earth resources technology satellite–1*. Washington, D.C.: V. I, Technical Presentations (NTIS order no. N7-328207); V. II, Summary of Results (NTIS order no. N73-28389); V. III (Discipline Summary Reports, NTIS order no. N73-28405).

—*Earth resources survey symposium*, Houston, Texas, June 1975.

—*Third Earth resources technology satellite–1 symposium*. Washington, D.C.: V. 1A, Technical Presentations (NTIS order no. N74-3075); V. IB, Technical Presentations (NTIS order no. N74-30774); V. II, Summary of Results (NTIS order no. 10549).

National Technical Information Service. *NASA Earth Resources Survey Program Weekly Abstracts*. Weekly list of documents related to the Earth Resources programs available through NTIS. Also lists Landsat experiment reports as they become available. 5285 Port Royal Road, Springfield, Virginia 22161.

Patterson, C. L. Special issue on interactive digital image processing. *Computer*, 1977, *10*(8).

Proceedings, IEEE Computer Society Conference on Pattern Recognition and Image Processing, 6–8 June 1977, Troy, New York (IEEE Publ. 77CH 1208-9 C). Long Beach, California: IEEE Computer Society, 1977.

Proceedings of the Workshop on Picture Data Description and Management, 21–22 April 1977,

Chicago, Illinois. (IEEE Publ. 77CH 1187-4 C). Long Beach, California: IEEE Computer Society, 1977.

Reeves, R. G. (Ed.). *Manual of remote sensing*. Falls Church, Virginia: American Society of Photogrammetry, 1975.

Rosenfeld, A. *Picture processing by computer*. New York: Academic Press, 1969.

Rosenfeld, A. Picture processing: 1976. *Computer Graphics Image Processing*, 1977, *6*, 157–183.

Rosenfeld, A., & Kak, A. C., *Digital picture processing*. New York: Academic Press, 1976.

Rudd, R. D. *Remote sensing: A better view*. North Scituate, Mass.: Duxbury Press, 1974.

Scherz, J. P. *An introduction to remote sensing for environmental monitoring*. Madison: University of Wisconsin, 1970.

Short, N. M. *Earth observations from space: Outlook for geological sciences* (NTIS order no. N74-12157). Greenbelt, Md.: National Aeronautics and Space Administration.

Smith, J. T., Jr. (Ed.). *Manual of color aerial photography*. Falls Church, Va.: American Society of Photogrammetry, 1968.

Smith, W. L. Remote sensing applications for mineral exploration. Stroudsburg, Pennsylvania: Dowden, Hutchinson and Ross, 1977.

Technology Application Center (TAC). *Quarterly Literature Review of the Remote Sensing of Natural Resources*. Most complete of the current on-going bibliographies. Lists remote-sensing-related documents published by NASA, NTIS, Engineering Index, and four other databases. It is subdivided and covers all areas of remote sensing applications and research. University of New Mexico, Albuquerque.

Ullmann, J. R., & Rosenfeld, A. Picture recognition and analysis. *The Radio and Electronic Engineer*, 1977, *47*, 33–48.

Way, D. *Terrain analysis: A guide to site selection using aerial photographic interpretation*. Stroudsburg, Pa.: Dowden Hutchinson, Ross, 1973.

Wiedel, J. W., & Kleckner, R. *Using remote sensor data for land use mapping and inventory: A user guide*. U.S. Geol. Survey IR-253, NTIS order no. PB242-813) Washington, D.C.: U.S. Gov't. Printing Office.

Wilde, C. O., & Barrett, E. (Eds.). *Image Science Mathematics*, Western Periodicals Corp. (Conference held 10–12 November 1976, Monterey, California).

Williams, R. S., & Carter, W. D. (Eds.) *ERTS-1: A new window on our planet*. (U.S. Geol. Survey Prof. Paper 929). Washington, D.C.: U.S. Gov't Printing Office.

SOURCES OF RELATED INFORMATION

The U.S. Department of Agriculture, U.S. Department of Commerce, U.S. Department of the Interior, and the National Aeronautics and Space Administration (NASA) maintain libraries containing reports on various aspects of Earth resources applications and technology. The National Referral Center of the Library of Congress and the Smithsonian Science Information Exchange offer specific information on research projects.

National Referral Center, Library of Congress

The National Referral Center functions as an information desk, directing those who have questions concerning scientific and technical subjects to organizations or individuals who have specialized knowledge. It provides names, addresses, telephone numbers, and brief descriptions of appropriate information sources: professional societies, university research bureaus, and institutes, federal and state agencies, industrial laboratories, museum specimen collections, testing stations, technical libraries, information and document centers, abstracting and indexing services, and individual experts who are willing to share their knowledge with others.

Services are available without charge. Requests may be made by writing to the Library of Congress, Science and Technology Division, National Referral Center, Washington, D.C. 20540; or by visiting the center on the fifth floor of the Library of Congress Annex, Second and Independence Avenue, SE, Washington, D.C.

PUBLICATIONS

1. A booklet describing services of the National Referral Center is available without charge.
2. *Selected Information Resources on Remote Sensing.* List of 23 data collection and information centers and sources with brief descriptions of holdings or services, addresses, and telephone numbers. Available without charge.
3. *Physical Sciences, Engineering: A Directory of Information Resources in the United States.* Published in 1971. For sale from the Superintendent of Documents, U.S. Government Printing Office, Washington, D.C. 20402.

Smithsonian Science Information Exchange

The Smithsonian Science Information Exchange (SSIE) serves as a source of information on research in progress. It constantly updates a database of more than 100,000 current research projects funded by both public and private sources. Each record describes who supports the projects, who is doing the work, where and when the research is performed, and a technical summary of the project. The collection covers basic and applied research in life, physical, social, behavioral, and engineering sciences. The file can be searched by subject, organization, investigator, etc; the output is a packet of 1-page project summaries. Charges vary with the service. Requests may be made by writing to the Smithsonian Science Information Exchange, 1730 M Street, N.W., Washington, D.C. 20036.

PUBLICATIONS

1. *SSIE Description of Services.* Booklet available without charge.
2. *SSIE Science Newsletter.* Lists preselected information packages of current research summaries in areas of special interest.

Subject Index